BANACH ALGEBRAS
AND SEVERAL COMPLEX VARIABLES

Banach Algebras and Several Complex Variables

JOHN WERMER
Brown University

MARKHAM PUBLISHING COMPANY
CHICAGO

MARKHAM MATHEMATICS SERIES

William J. LeVeque, Editor

Anderson, *Graph Theory and Finite Combinatorics*
Cantrell and Edwards, eds., *Topology of Manifolds*
Gaal, *Classical Galois Theory with Examples*
Gioia, *The Theory of Numbers: An Introduction*
Greenspan, *Introduction to Numerical Analysis and Applications*
Hu, *Calculus*
Hu, *Cohomology Theory*
Hu, *Elementary Functions and Coordinate Geometry*
Hu, *Linear Algebra with Differential Equations*
Knopp, *Theory of Area*
Knopp, *Modular Functions in Analytic Number Theory*
Peterson, *Foundations of Algebra and Number Theory*
Stark, *An Introduction to Number Theory*

LECTURES IN ADVANCED MATHEMATICS

Davenport, *1. Multiplicative Number Theory*
Storer, *2. Cyclotomy and Difference Sets*
Engeler, *3. Formal Languages: Automata and Structures*
Garsia, *4. Topics in Almost Everywhere Convergence*

Printed in the United States of America
Library of Congress Catalog Card Number: 72-136624
Standard Book Number: 8410-1008-0

To Kerstin

PREFACE

The object of this book is to introduce the reader to a certain area of complex analysis that has flourished during the past fifteen years. It is the borderland between the theory of commutative Banach algebras and classical function theory in one or more complex variables.

We shall chiefly be concerned with questions involving functions of several complex variables.

Our aim has been to make the exposition elementary and self-contained. We have cheerfully sacrificed generality and completeness all along the way to make it easier for the reader to understand what is going on.

We assume that the reader is familiar with the basic Gelfand theory as exposed, for instance, on pages 8–28 of A. Browder's *Introduction to Function Algebras* (W. A. Benjamin, New York, 1969). Also, we assume knowledge of a few basic properties of Banach spaces.

Exercises of varying degrees of difficulty are included in the text and the reader should try to solve as many of these as he can. Solutions to starred exercises are given in Section 18.

In Sections 6 through 9 we follow the developments in Chapter 1 of R. Gunning and H. Rossi, *Analytic Functions of Several Complex Variables* (Prentice-Hall, Englewood Cliffs, N.J., 1965), or in Chapter III of L. Hörmander, *An Introduction to Complex Analysis in Several Variables* (Van Nostrand Reinhold, New York, 1966).

Our book could well be used as a companion volume to the book by Browder, which emphasizes relations between Banach algebras and problems in one complex variable.

T. W. Gamelin's *Uniform Algebras* (Prentice-Hall, Englewood Cliffs, N.J., 1969) has a good deal of overlap with the subject matter of our book but covers a very much wider range of material and proceeds at a considerably brisker pace. Gamelin's book and the books by Gunning and Rossi and by Hörmander are strongly recommended to a reader of this volume who wants to learn about adjacent areas of analysis.

I want to thank Richard Basener and John O'Connell, who read the manuscript and made many helpful mathematical suggestions and improvements. I am also very much indebted to my colleagues A. Browder and B. Weinstock for valuable comments. Warm thanks are due to Irving Glicksberg.

I am very grateful to Mrs. Roberta Weller for her excellent work in typing the manuscript.

JOHN WERMER

Providence, R.I.
March 1971

CONTENTS

1

PRELIMINARIES AND NOTATIONS

Let X be a compact Hausdorff space.

$C_R(X)$	is the space of all real-valued continuous functions on X.
$C(X)$	is the space of all complex-valued continuous functions on X.
	By a *measure* μ on X we shall mean a complex-valued Baire measure of finite total variation on X.
$\lvert\mu\rvert$	is the positive total variation measure corresponding to μ.
$\lVert\mu\rVert$	is $\lvert\mu\rvert(X)$.
$\mathbf{C}$	is the complex numbers.
$\mathbf{R}$	is the real numbers.
Z	is the integers.
$\mathbf{C}^n$	is the space of n-tuples of complex numbers.
	Fix n and let Ω be an open subset of $\mathbf{C}^n$.
$C^k(\Omega)$	is the space of k-times continuously differentiable functions on Ω, $k = 1,2,\ldots,\infty$.
$C_0^k(\Omega)$	is the subset of $C^k(\Omega)$ consisting of functions with compact support contained in Ω.
$H(\Omega)$	is the space of holomorphic functions defined on Ω.
	By *Banach algebra* we shall mean a commutative Banach algebra with unit. Let $\mathfrak{A}$ be such an object.
$\mathcal{M}(\mathfrak{A})$	is the space of maximal ideals of $\mathfrak{A}$. When no ambiguity arises, we shall write $\mathcal{M}$ for $\mathcal{M}(\mathfrak{A})$. If m is a homomorphism of $\mathfrak{A} \to \mathbf{C}$, we shall frequently identify m with its kernel and regard m as an element of $\mathcal{M}$.
	For f in $\mathfrak{A}$, M in $\mathcal{M}$,
$\hat{f}(M)$	is the value at f of the homomorphism of $\mathfrak{A}$ into $\mathbf{C}$ corresponding to M. We shall sometimes write $f(M)$ instead of $\hat{f}(M)$.
$\hat{\mathfrak{A}}$	is the algebra consisting of all functions $\hat{f}$ on $\mathcal{M}$ with f in $\mathfrak{A}$. For x in $\mathfrak{A}$,
$\sigma(x)$	is the spectrum of $x = \{\lambda \in \mathbf{C} \mid \lambda - x$ has no inverse in $\mathfrak{A}\}$.
rad $\mathfrak{A}$	is the radical of $\mathfrak{A}$.

For $z = (z_1, \ldots, z_n) \in \mathbf{C}^n$,
$|z|$ $= \sqrt{|z_1|^2 + |z_2|^2 + \cdots + |z_n|^2}$.
For S a subset of a topological space,
$\mathring{S}$ is the interior of S,
$\bar{S}$ is the closure of S, and
∂S is the boundary of S.
For X a compact subset of $\mathbf{C}^n$,
$P(X)$ is the closure in $C(X)$ of the polynomials in the coordinates.
Let Ω be a plane region with compact closure $\bar{\Omega}$. Then
$A(\Omega)$ is the algebra of all functions continuous on $\bar{\Omega}$ and holomorphic on Ω.

Let X be a compact space, $\mathscr{L}$ a subset of $C(X)$, and μ a measure on X. We write $\mu \perp \mathscr{L}$ and say μ is *orthogonal* to $\mathscr{L}$ if

$$\int f\, d\mu = 0 \qquad \text{for all } f \text{ in } \mathscr{L}.$$

We shall frequently use the following result (or its real analogue) without explicitly appealing to it:

THEOREM (RIESZ–BANACH)

Let $\mathscr{L}$ be a linear subspace of $C(X)$ and fix g in $C(X)$. If for every measure μ on X

$$\mu \perp \mathscr{L} \text{ implies } \mu \perp g,$$

then g lies in the closure of $\mathscr{L}$. In particular, if

$$\mu \perp \mathscr{L} \text{ implies } \mu = 0,$$

then $\mathscr{L}$ is dense in $C(X)$.

We shall need the following elementary fact, left to the reader as

Exercise 1.1. Let X be a compact space. Then to every maximal ideal M of $C(X)$ corresponds a point x_0 in X such that $M = \{f \text{ in } C(X) | f(x_0) = 0\}$. Thus $\mathscr{M}(C(X)) = X$.

Here are some examples of Banach algebras.

(a) Let T be a bounded linear operator on a Hilbert space H and let $\mathfrak{A}$ be the closure in operator norm on H of all polynomials in T. Impose the operator norm on $\mathfrak{A}$.

(b) Let $C^1(a, b)$ denote the algebra of all continuously differentiable functions on the interval $[a, b]$, with

$$\|f\| = \max_{[a,b]} |f| + \max_{[a,b]} |f'|.$$

(c) Let Ω be a plane region with compact closure $\bar{\Omega}$. Let $A(\Omega)$ denote the algebra of all functions continuous on $\bar{\Omega}$ and holomorphic in Ω, with

$$\|f\| = \max_{z \in \bar{\Omega}} |f(z)|.$$

(d) Let X be a compact subset of C^n. Denote by $P(X)$ the algebra of all functions defined on X which can be approximated by polynomials in the coordinates $z_1, \ldots, z_n$

uniformly on X, with

$$\|f\| = \max_X |f|.$$

(e) Denote by $H^\infty(D)$ the algebra of all bounded holomorphic functions defined in the open unit disk D. Put

$$\|f\| = \sup_D |f|.$$

(f) Let X be a compact subset of the plane. $R(X)$ denotes the algebra of all functions on X which can be uniformly approximated on X by functions holomorphic in some neighborhood of X. Take

$$\|f\| = \max_X |f|.$$

(g) Let X be a compact Hausdorff space. On the algebra $C(X)$ of all complex-valued continuous functions on X we impose the norm

$$\|f\| = \max_X |f|.$$

Definition. Let X be a compact Hausdorff space. A *uniform algebra* on X is an algebra $\mathfrak{A}$ of continuous complex-valued functions on X satisfying

(i) $\mathfrak{A}$ is closed under uniform convergence on X.
(ii) $\mathfrak{A}$ contains the constants.
(iii) $\mathfrak{A}$ separates the points of X.

$\mathfrak{A}$ is normed by $\|f\| = \max_X |f|$ and so becomes a Banach algebra.

Note that $C(X)$ is a uniform algebra on X, and that every other uniform algebra on X is a proper closed subalgebra of $C(X)$. Among our examples, (c), (d), (f), and (g) are uniform algebras; (a) is not, except for certain T, and (b) is not.

If $\mathfrak{A}$ is a uniform algebra, then clearly

(1) $$\|x^2\| = \|x\|^2 \qquad \text{for all } x \in \mathfrak{A}.$$

Conversely, let $\mathfrak{A}$ be a Banach algebra satisfying (1). We claim that $\mathfrak{A}$ is isometrically isomorphic to a uniform algebra. For (1) implies that

$$\|x^4\| = \|x\|^4, \ldots, \|x^{2^n}\| = \|x\|^{2^n}, \qquad \text{all } n.$$

Hence

$$\|x\| = \lim_{k\to\infty} \|x^k\|^{1/k} = \max_{\mathcal{M}} |\hat{x}|.$$

Since $\mathfrak{A}$ is complete in its norm, it follows that $\hat{\mathfrak{A}}$ is complete in the uniform norm on $\mathcal{M}$, so $\hat{\mathfrak{A}}$ is closed under uniform convergence on $\mathcal{M}$. Hence $\hat{\mathfrak{A}}$ is a uniform algebra on $\mathcal{M}$ and the map $x \to \hat{x}$ is an isometric isomorphism from $\mathfrak{A}$ to $\hat{\mathfrak{A}}$.

It follows that the algebra $H^\infty(D)$ of example (e) is isometrically isomorphic to a uniform algebra on a suitable compact space.

In the later portions of this book, starting with Section 10, we shall study uniform algebras, whereas the earlier sections (as well as Section 15) will be concerned with arbitrary Banach algebras.

Throughout, when studying general theorems, the reader should keep in mind some concrete examples such as those listed under (a) through (g), and he should make clear to himself what the general theory means for the particular examples.

Exercise 1.2. Let $\mathfrak{A}$ be a uniform algebra on X and let h be a homomorphism of $\mathfrak{A} \to \mathbf{C}$. Show that there exists a probability measure (positive measure of total mass 1) μ on X so that

$$h(f) = \int_X f \, d\mu, \qquad \text{all } f \text{ in } \mathfrak{A}.$$

2

CLASSICAL APPROXIMATION THEOREMS

Let X be a compact Hausdorff space. Let $\mathfrak{A}$ be a subalgebra of $C_R(X)$ which contains the constants.

THEOREM 2.1 (REAL STONE–WEIERSTRASS THEOREM)

If $\mathfrak{A}$ separates the points of X, then $\mathfrak{A}$ is dense in $C_R(X)$.

We shall deduce this result from the following general theorem:

PROPOSITION 2.2

Let B be a real Banach space and B^* its dual space taken in the weak- topology. Let K be a nonempty compact convex subset of B^*. Then K has an extreme point.*

Note. If W is a real vector space, S a subset of W, and p a point of S, then p is called an *extreme point* of S provided

$$p = \tfrac{1}{2}(p_1 + p_2), \qquad p_1, p_2 \in S \Rightarrow p_1 = p_2 = p.$$

If S is a convex set and p an extreme point of S, then $0 < \theta < 1$ and $p = \theta p_1 + (1 - \theta)p_2$ implies that $p_1 = p_2 = p$.

We shall give the proof for the case that B is separable.

Proof. Let $\{L_n\}$ be a countable dense subset of B. If $y \in B^*$, put

$$L_n(y) = y(L_n).$$

Define

$$l_1 = \sup_{x \in K} L_1(x).$$

Since K is compact and L_1 continuous, l_1 is finite and attained; i.e., $\exists x_1 \in K$ with $L_1(x_1) = l_1$. Put

$$l_2 = \sup L_2(x) \text{ over all } x \in K, \qquad \text{with } L_1(x) = l_1.$$

Again, the sup is taken over a compact set, contained in K, so $\exists x_2 \in K$ with

$$L_2(x_2) = l_2 \qquad \text{and} \qquad L_1(x_2) = l_1.$$

Going on in this way, we get a sequence $x_1, x_2, \ldots$ in K so that for each n,

$$L_1(x_n) = l_1, L_2(x_n) = l_2, \ldots, L_n(x_n) = l_n,$$

and

$$l_{n+1} = \sup L_{n+1}(x) \text{ over } x \in K \qquad \text{with } L_1(x) = l_1, \ldots, L_n(x) = l_n.$$

Let x^* be an accumulation point of $\{x_n\}$. Then $x^* \in K$.

$L_j(x_n) = l_j$ for all large n. So $L_j(x^*) = l_j$ for all j.

We claim that x^* is an extreme point in K. For let

$$x^* = \tfrac{1}{2}y_1 + \tfrac{1}{2}y_2, \qquad y_1, y_2 \in K.$$

$l_1 = L_1(x^*) = \frac{1}{2}L_1(y_1) + \frac{1}{2}L_1(y_2)$. Since $L_1(y_j) \leq l_1$, $j = 1, 2$, $L_1(y_1) = L_1(y_2) = l_1$. Also,

$$l_2 = L_2(x^*) = \tfrac{1}{2}L_2(y_1) + \tfrac{1}{2}L_2(y_2).$$

Since $L_1(y_1) = l_1$ and $y_1 \in K$, $L_2(y_1) \leq l_2$. Similarly, $L_2(y_2) \leq l_2$. Hence

$$L_2(y_1) = L_2(y_2) = l_2.$$

Proceeding in this way, we get

$$L_k(y_1) = L_k(y_2) \qquad \text{for all } k.$$

But $\{L_k\}$ was dense in B. It follows that $y_1 = y_2$. Thus x^* is extreme in K. Q.E.D.

Note. Proposition 2.2 (without separability assumption) is proved in [4, pp. 439–40]. In the application of Proposition 2.2 to the proof of Theorem 2.1 (see below), $C_R(X)$ is separable provided X is a metric space.

Proof of Theorem 2.1. Let

$$K = \{\mu \in (C_R(X))^* | \mu \perp \mathfrak{A} \text{ and } \|\mu\| \leq 1\}.$$

K is a compact, convex set in $(C_R(X))^*$. (Why?) Hence K has an extreme point σ, by Proposition 2.2. Unless $K = \{0\}$, we can choose σ with $\|\sigma\| = 1$. Since $1 \in \mathfrak{A}$ and so

$$\int 1 \, d\sigma = 0,$$

σ cannot be a point mass and so $\exists$ distinct points x_1 and x_2 in the carrier of σ.

Choose $g \in \mathfrak{A}$ with $g(x_1) \neq g(x_2)$, $0 < g < 1$. (How?) Then

$$\sigma = g \cdot \sigma + (1 - g)\sigma = \|g\sigma\| \frac{g\sigma}{\|g\sigma\|} + \|(1 - g)\sigma\| \frac{(1 - g)\sigma}{\|(1 - g)\sigma\|}.$$

Also,

$$\|g\sigma\| + \|(1-g)\sigma\| = \int g\,d|\sigma| + \int (1-g)\,d|\sigma| = \int d|\sigma| = \|\sigma\| = 1.$$

Thus σ is a convex combination of $g\sigma/\|g\sigma\|$ and $(1-g)\sigma/\|(1-g)\sigma\|$. But both of these measures lie in K. (Why?) Hence

$$\sigma = \frac{g\sigma}{\|g\sigma\|}.$$

It follows that g is constant a.e. - $d|\sigma|$. But $g(x_1) \neq g(x_2)$ and g is continuous which gives a contradiction.

Hence $K = \{0\}$ and so $\mu \in (C_R(X))^*$ and $\mu \perp \mathfrak{A} \Rightarrow \mu = 0$. Thus $\mathfrak{A}$ is dense in $C_R(X)$, as claimed.

THEOREM 2.3 (COMPLEX STONE–WEIERSTRASS THEOREM)

$\mathfrak{A}$ is a subalgebra of $C(X)$ containing the constants and separating points. If

(1) $$f \in \mathfrak{A} \Rightarrow \bar{f} \in \mathfrak{A},$$

then $\mathfrak{A}$ is dense in $C(X)$.

Proof. Let $\mathscr{L}$ consists of all real-valued functions in $\mathfrak{A}$. Since by (1) $\mathscr{L}$ contains Re f and Im f for each $f \in \mathfrak{A}$, $\mathscr{L}$ separates points on X. Evidently $\mathscr{L}$ is a subalgebra of $C_R(X)$ containing the (real) constants. By Theorem 2.1 $\mathscr{L}$ is then dense in $C_R(X)$. It follows that $\mathfrak{A}$ is dense in $C(X)$. (How?)

Let Σ_R denote the real subspace of $C^n = \{(z_1, \ldots, z_n) \in C^n | z_j \text{ is real, all } j\}$.

COROLLARY 1

Let X be a compact subset of Σ_R. Then $P(X) = C(X)$.

Proof. Let $\mathfrak{A}$ be the algebra of all polynomials in $z_1, \ldots, z_n$ restricted to X. $\mathfrak{A}$ then satisfies the hypothesis of the last theorem, and so $\mathfrak{A}$ is dense in $C(X)$; i.e., $P(X) = C(X)$.

COROLLARY 2

Let I be an interval on the real line. Then $P(I) = C(I)$.

This is, of course, the Weierstrass approximation theorem (slightly complexified).

Let us replace I by an arbitrary compact subset X of C. When does $P(X) = C(X)$? It is easy to find necessary conditions on X. (Find some.) However, to get a complete solution, some machinery must first be built up.

The machinery we shall use will be some elementary potential theory for the Laplace operator Δ in the plane, as well as for the Cauchy–Riemann operator

$$\frac{\partial}{\partial \bar{z}} = \frac{1}{2}\left(\frac{\partial}{\partial x} + i\frac{\partial}{\partial y}\right).$$

These general results will then be applied to several approximation problems in the plane, including the above problem of characterizing those X for which $P(X) = C(X)$.

Let μ be a measure of compact support $\subset \mathbf{C}$. We define the *logarithmic potential* μ^* of μ by

$$\mu^*(z) = \int \log\left|\frac{1}{z-\zeta}\right| d\mu(\zeta). \tag{2}$$

We define the *Cauchy transform* $\hat{\mu}$ of μ by

$$\hat{\mu}(z) = \int \frac{1}{\zeta - z} d\mu(\zeta). \tag{3}$$

LEMMA 2.4

The functions

$$\int \left|\log\left|\frac{1}{z-\zeta}\right|\right| d|\mu|(\zeta) \qquad \textit{and} \qquad \int \left|\frac{1}{\zeta - z}\right| d|\mu|(\zeta)$$

are summable - $dx\,dy$ *over compact sets in* $\mathbf{C}$. *It follows that these functions are finite* a.e. - $dx\,dy$ *and hence that* μ^* *and* $\hat{\mu}$ *are defined* a.e. - $dx\,dy$.

Since $1/r \geq |\log r|$ for small $r > 0$, we need only consider the second integral. Fix $R > 0$ with $\operatorname{supp}|\mu| \subset \{z \,|\, |z| < R\}$.

$$\gamma = \int_{|z| \leq R} dx\,dy \left\{\int \left|\frac{1}{\zeta - z}\right| d|\mu|(\zeta)\right\} = \int d|\mu|(\zeta) \int_{|z| \leq R} \frac{dx\,dy}{|z - \zeta|}.$$

For $\zeta \in \operatorname{supp}|\mu|$ and $|z| \leq R$, $|z - \zeta| \leq 2R$.

$$\int_{|z| \leq R} \frac{dx\,dy}{|z-\zeta|} \leq \int_{|z'| \leq 2R} \frac{dx'\,dy'}{|z'|} = \int_0^{2R} r\,dr \int_0^{2\pi} \frac{d\theta}{r} = 4\pi R.$$

Hence $\gamma \leq 4\pi R \cdot \|\mu\|$. Q.E.D.

LEMMA 2.5

Let $F \in C_0^1(\mathbf{C})$. *Then*

$$F(\zeta) = -\frac{1}{\pi} \iint_{\mathbf{C}} \frac{\partial F}{\partial \bar{z}} \frac{dx\,dy}{z - \zeta}, \qquad \text{all } \zeta \in C. \tag{4}$$

Note. The proof uses differential forms. If this bothers you, read the proof after reading Sections 4 and 5, where such forms are discussed, or make up your own proof.

Proof. Fix ζ and choose $R > |\zeta|$ with $\operatorname{supp} F \subset \{z \,|\, |z| < R\}$. Fix $\varepsilon > 0$ and small. Put $\Omega_\varepsilon = \{z \,|\, |z| < R \text{ and } |z - \zeta| > \varepsilon\}$.

The 1-form $F\,dz/z - \zeta$ is smooth on Ω_ε and

$$d\left(\frac{F\,dz}{z-\zeta}\right) = \frac{\partial}{\partial \bar{z}}\left(\frac{F}{z-\zeta}\right) d\bar{z} \wedge dz = \frac{\partial F}{\partial \bar{z}} \frac{d\bar{z} \wedge dz}{z - \zeta}.$$

By Stokes's theorem

$$\int_{\Omega_\varepsilon} d\left(\frac{F\,dz}{z-\zeta}\right) = \int_{\partial\Omega_\varepsilon} \frac{F\,dz}{z-\zeta}.$$

Since $F = 0$ on $\{z \| z| = R\}$, the right side is

$$\int_{|z-\zeta|=\varepsilon} \frac{F\,dz}{z-\zeta} = -\int_0^{2\pi} F(\zeta + \varepsilon e^{i\theta}) i\,d\theta,$$

so

$$\int_{\Omega_\varepsilon} \frac{\partial F}{\partial \bar{z}} \frac{d\bar{z} \wedge dz}{z-\zeta} = -\int_0^{2\pi} F(\zeta + \varepsilon e^{i\theta}) i\,d\theta.$$

Letting $\varepsilon \to 0$ we get

$$\int_{|z|<R} \frac{\partial F}{\partial \bar{z}} \frac{d\bar{z} \wedge dz}{z-\zeta} = -2\pi i F(\zeta).$$

Since $\partial F/\partial \bar{z} = 0$ for $|z| > R$ and since $d\bar{z} \wedge dz = 2i\,dx \wedge dy$, this gives

$$\int \frac{\partial F}{\partial \bar{z}} \frac{dx\,dy}{z-\zeta} = -\pi F(\zeta),$$

i.e., (4).

Note. The intuitive content of (4) is that arbitrary smooth functions can be synthesized from functions

$$f_\lambda(\zeta) = \frac{1}{\lambda - \zeta}$$

by taking linear combinations and then limits.

LEMMA 2.6

Let $G \in C_0^2(\mathbf{C})$. *Then*

$$G(\zeta) = -\frac{1}{2\pi} \iint_{\mathbf{C}} \Delta G(z) \log \frac{1}{|z-\zeta|}\,dx\,dy, \qquad \text{all } \zeta \in \mathbf{C}. \tag{5}$$

Proof. The proof is very much like that of Lemma 2.5. With Ω_ε as in that proof, start with Green's formula

$$\iint_{\Omega_\varepsilon} (u\Delta v - v\Delta u)\,dx\,dy = \int_{\partial\Omega_\varepsilon} \left(u\frac{\partial v}{\partial n} - v\frac{\partial u}{\partial n}\right) ds$$

and take $u = G$, $v = \log|z-\zeta|$. We leave the details to you.

LEMMA 2.7

If μ *is a measure with compact support in* $\mathbf{C}$. *and if* $\hat{\mu}(z) = 0$ a.e. - $dx\,dy$, *then* $\mu = 0$. *Also if* $\mu^*(z) = 0$ a.e. - $dx\,dy$, *then* $\mu = 0$.

Proof. Fix $g \in C_0^1(\mathbf{C})$. By (4)

$$\int g(\zeta)\, d\mu(\zeta) = \int d\mu(\zeta)\left[-\frac{1}{\pi}\int \frac{\partial g}{\partial \bar{z}}(z)\frac{dx\, dy}{z-\zeta}\right].$$

Fubini's theorem now gives

$$\frac{1}{\pi}\int \frac{\partial g}{\partial \bar{z}}(z)\hat{\mu}(z)\, dx\, dy = \int g\, d\mu. \tag{6}$$

Since $\hat{\mu} = 0$ a.e., we deduce that

$$\int g\, d\mu = 0.$$

But the class of functions obtained by restricting to supp μ the functions in $C_0^1(\mathbf{C})$ is dense in $C(\text{supp}\ \mu)$ by the Stone–Weierstrass theorem. Hence $\mu = 0$.

Using (5), we get similarly for $g \in C_0^2(\mathbf{C})$,

$$-\int g\, d\mu = \frac{1}{2\pi}\int \Delta g(z)\cdot \mu^*(z)\, dx\, dy$$

and conclude that $\mu = 0$ if $\mu^* = 0$ a.e.

As a first application, consider a compact set $X \subset C$.

THEOREM 2.8 (HARTOGS–ROSENTHAL)

Assume that X has Lebesgue two-dimensional measure 0. *Then rational functions whose poles lie off X are uniformly dense in C(X).*

Proof. Let W be the linear space consisting of all rational functions holomorphic on X. W is a subspace of $C(X)$. To show W dense, we consider a measure μ on X with $\mu \perp W$. Then $\hat{\mu}(z) = \int d\mu(\zeta)/\zeta - z = 0$ for $z \notin X$, since $1/\zeta - z \in W$ for such z, and $\mu \perp W$.

Since X has measure 0, $\hat{\mu} = 0$ a.e. - $dx\, dy$. Lemma 2.7 yields $\mu = 0$.

Hence $\mu \perp W \Rightarrow \mu = 0$ and so W is dense. Q.E.D.

As a second application, consider an open set $\Omega \subset \mathbf{C}$ and a compact set $K \subset \Omega$. (In the proofs of the next two theorems we shall suppose Ω bounded and leave the modifications for the general case to the reader.)

THEOREM 2.9 (RUNGE)

If F is a holomorphic function defined on Ω, *there exists a sequence* $\{R_n\}$ *of rational functions holomorphic in* Ω *with*

$$R_n \to F \text{ uniformly on } K.$$

Proof. Let $\Omega_1, \Omega_2, \ldots$ be the components of $\mathbf{C}\setminus K$. It is no loss of generality to assume that each Ω_j meets the complement of Ω. (Why?) Fix $p_j \in \Omega_j \setminus \Omega$.

Let W be the space of all rational functions regular except for possible poles at some of the p_j, restricted to K. Then W is a subspace of $C(K)$ and it suffices to show that W contains F in its closure.

Choose a measure μ on K with $\mu \perp W$. We must show that $\mu \perp F$.

Fix $\phi \in C^\infty(\mathbf{C})$, supp $\phi \subset \Omega$ and $\phi = 1$ in a neighborhood N of K.

Using (6) with $g = F \cdot \phi$ we get

$$\frac{1}{\pi}\int \frac{\partial(F\phi)}{\partial \bar{z}}(z)\hat{\mu}(z)\, dx\, dy = \int F\phi\, d\mu. \tag{7}$$

Fix j.

$$\hat{\mu}(z) = \int \frac{d\mu(\zeta)}{\zeta - z}$$

is analytic in Ω_j and

$$\frac{d^k\hat{\mu}}{dz^k}(p_j) = k! \int \frac{d\mu(\zeta)}{(\zeta - p_j)^{k+1}}, \qquad k = 0, 1, 2, \ldots.$$

The right-hand side is 0 since $(\zeta - p_j)^{-(k+1)} \in W$ and $\mu \perp W$. Thus all derivatives of $\hat{\mu}$ vanish at p_j and hence $\hat{\mu} = 0$ in Ω_j. Thus $\hat{\mu} = 0$ on $\mathbf{C} \setminus K$. Also, $F\phi = F$ is analytic in N, and so

$$\frac{\partial}{\partial \bar{z}}(F\phi) = 0 \text{ on } K.$$

The integrand on the left in (7) thus vanishes everywhere, and so

$$\int F\, d\mu = \int F\phi\, d\mu = 0.$$

Thus $\mu \perp W \Rightarrow \mu \perp F$. Q.E.D.

When can we replace "rational function" by "polynomial" in the last theorem?

Suppose that Ω is *multiply connected*. Then we cannot.

The reason is this: We can choose a simple closed curve β lying in Ω such that some point z_0 in the interior of β lies outside Ω. Put

$$F(z) = \frac{1}{z - z_0}.$$

Then F is holomorphic in Ω. Suppose that $\exists$ a sequence of polynomials $\{P_n\}$ converging uniformly to F on β. Then

$$(z - z_0)P_n - 1 \to 0 \text{ uniformly on } \beta.$$

By the maximum principle

$$(z - z_0)P_n - 1 \to 0 \text{ inside } \beta.$$

But this is false for $z = z_0$.

THEOREM 2.10 (RUNGE)

Let Ω be a simply connected region and fix G holomorphic in Ω. If K is a compact subset of Ω, then $\exists$ a sequence $\{P_n\}$ of polynomials converging uniformly to G on K.

Proof. Without loss of generality we may assume that $\mathbf{C} \setminus K$ is connected.

Fix a point p in $\mathbf{C}$ lying outside a disk $\{z \| z| \leq R\}$ which contains K. The proof of the last theorem shows that $\exists$ rational functions R_n with sole pole at p with

$$R_n \to G \text{ uniformly on } K.$$

The Taylor expansion around 0 for R_n converges uniformly on K. Hence we can replace R_n by a suitable partial sum P_n of this Taylor series, getting

$$P_n \to G \text{ uniformly on } K. \qquad \text{Q.E.D.}$$

We return now to the problem of describing those compact sets X in the z-plane which satisfy $P(X) = C(X)$.

Let p be an interior point of X. Then every f in $P(X)$ is analytic at p. Hence the condition

(8) The interior of X is empty.

is necessary for $P(X) = C(X)$.

Let Ω_1 be a bounded component of $\mathbf{C} \setminus X$. Fix $F \in P(X)$. Choose polynomials P_n with

$$P_n \to F \text{ uniformly on } X.$$

Since $\partial\Omega_1 \subset X$,

$$|P_n - P_m| \to 0 \text{ uniformly on } \partial\Omega_1$$

as $n, m \to 0$. Hence by the maximum principle

$$|P_n - P_m| \to 0 \text{ uniformly on } \Omega_1.$$

Hence P_n converges uniformly on $\Omega_1 \cup \partial\Omega_1$ to a function holomorphic on Ω_1, continuous on $\Omega_1 \cup \partial\Omega_1$, and $= F$ on $\partial\Omega_1$.

This restricts the elements F of $P(X)$ to a proper subset of $C(X)$. (Why?) Hence the condition

(9) $\mathbf{C} \setminus X$ is connected.

is also necessary for $P(X) = C(X)$.

THEOREM 2.11 (LAVRENTIEFF)

If (8) *and* (9) *hold, then* $P(X) = C(X)$.

Note that the Stone-Weierstrass theorem gives us no help here, for to apply it we should need to know that $\bar{z} \in P(X)$, and to prove that is as hard as the whole theorem.

The chief step in our proof is the demonstration of a certain continuity property of the logarithmic potential α^* of a measure α supported on a compact plane set E with connected complement, as we approach a boundary point z_0 of E from $\mathbf{C} \setminus E$.

LEMMA 2.12 (CARLESON)

Let E be a compact plane set with $\mathbf{C}\setminus E$ connected and fix $z_0 \in \partial E$. Then $\exists$ probability measures σ_t for each $t > 0$ with σ_t carried on $\mathbf{C}\setminus E$ such that:

Let α be a real measure on E satisfying

(10) $$\int_E \left|\log\left|\frac{1}{z_0 - \zeta}\right|\right| d|\alpha|(\zeta) < \infty.$$

Then

$$\lim_{t\to 0} \int \alpha^* \, d\sigma_t(z) = \alpha^*(z_0).$$

Proof. We may assume that $z_0 = 0$. Fix $t > 0$. Since $0 \in \partial E$ and $\mathbf{C}\setminus E$ is connected, $\exists$ a probability measure σ_t carried on $\mathbf{C}\setminus E$ such that

$$\sigma_t\{z | r_1 < |z| < r_2\} = \frac{1}{t}(r_2 - r_1) \qquad \text{for } 0 < r_1 < r_2 \le t$$

and $\sigma_t = 0$ outside $|z| \le t$.

If some line segment with 0 as one end point and length t happens to lie in $\mathbf{C}\setminus E$, we may of course take σ_t as $1/t \cdot$ linear measure on that segment. (In the general case, construct σ_t.)

Then for all $\zeta \in \mathbf{C}$ we have

$$\int \log\left|\frac{1}{z - \zeta}\right| d\sigma_t(z) \le \int \log\left|\frac{1}{|z| - |\zeta|}\right| d\sigma_t(z)$$

$$= \frac{1}{t}\int_0^t \log\frac{1}{|r - |\zeta||}\, dr = \log\frac{1}{|\zeta|} + \frac{1}{t}\int_0^t \log\frac{1}{|1 - r/|\zeta||}\, dr.$$

The last term is bounded above by a constant A independent of t and $|\zeta|$. (Why?) Hence we have

(11) $$\int \log\left|\frac{1}{z - \zeta}\right| d\sigma_t(z) \le \log\frac{1}{|\zeta|} + A, \qquad \text{all } \zeta, \text{ all } t > 0.$$

Also, as $t \to 0$, $\sigma_t \to$ point mass at 0. Hence for each fixed $\zeta \ne 0$,

(12) $$\int \log\left|\frac{1}{z - \zeta}\right| d\sigma_t(z) \to \log\frac{1}{|\zeta|}.$$

Now for fixed t Fubini's theorem gives

(13) $$\int \alpha^*(z)\, d\sigma_t(z) = \int \left\{ \int \log\left|\frac{1}{z - \zeta}\right| d\sigma_t(z) \right\} d\alpha(\zeta).$$

By (11), (12), and (10), the integrand on the right tends to $\log 1/|\zeta|$ dominatedly with respect to $|\alpha|$. Hence the right side approaches

$$\int \log\frac{1}{|\zeta|}\, d\alpha(\zeta) = \alpha^*(0)$$

as $t \to 0$, and so

$$\lim_{t \to 0} \int \alpha^*(z)\, d\sigma_t(z) = \alpha^*(0). \qquad \text{Q.E.D.}$$

Proof of Theorem 2.11. Let α be a real measure on X with $\alpha \perp \operatorname{Re}(P(X))$. Then

$$\int \operatorname{Re} \zeta^n \, d\alpha(\zeta) = 0, \qquad n \geq 0$$

and

$$\int \operatorname{Im} \zeta^n \, d\alpha = \int \operatorname{Re}(-i\zeta^n)\, d\alpha = 0, \qquad n \geq 0,$$

so that

$$\int \zeta^n \, d\alpha = 0, \qquad n \geq 0.$$

For $|z|$ large,

$$\log\left(1 - \frac{\zeta}{z}\right) = \sum_0^\infty c_n(z)\zeta^n,$$

the series converging uniformly for $\zeta \in X$. Hence

$$\int \log\left(1 - \frac{\zeta}{z}\right) d\alpha(\zeta) = \sum_0^\infty c_n(z) \int \zeta^n \, d\alpha(\zeta) = 0,$$

whence

$$\int \operatorname{Re}\left(\log\left(1 - \frac{\zeta}{z}\right)\right) d\alpha(\zeta) = 0$$

or

$$\int \log|z - \zeta| \, d\alpha(\zeta) - \int \log|z| \, d\alpha(\zeta) = 0,$$

whence

$$\int \log|z - \zeta| \, d\alpha(\zeta) = 0,$$

since $\alpha \perp 1$. Since

$$\int \log|z - \zeta| \, d\alpha(\zeta)$$

is harmonic in $\mathbf{C} \setminus X$, the function vanishes not only for large $|z|$, but in fact for all z in $\mathbf{C} \setminus X$, and so

$$\alpha^*(z) = 0, \qquad z \in \mathbf{C} \setminus X.$$

By Lemma 2.12 it follows that we also have

$$\alpha^*(z_0) = 0, \qquad z_0 \in X,$$

provided (10) holds at z^0. By Lemma 2.4 this implies that

$$\alpha^* = 0 \text{ a.e. - } dx\,dy.$$

By Lemma 2.7 this implies that $\alpha = 0$. Hence

(14) $$\operatorname{Re} P(X) \text{ is dense in } C_R(X).$$

Now choose $\mu \in P(X)^{\perp}$. Fix $z_0 \in X$ with

(15) $$\int \left| \frac{1}{z - z_0} \right| d|\mu|(z) < \infty.$$

Because of (14) we can find for each positive integer k a polynomial P_k such that

(16) $$|\operatorname{Re} P_k(z) - |z - z_0|| \leq \frac{1}{k}, \qquad z \in X$$

and

(17) $$P_k(z_0) = 0.$$

$$f_k(z) = \frac{e^{-kP_k(z)} - 1}{z - z_0}$$

is an entire function and hence its restriction to X lies in $P(X)$. Hence

(18) $$\int f_k \, d\mu = 0.$$

Equation (16) gives

$$\operatorname{Re} kP_k(z) - k|z - z_0| \geq -1,$$

whence

$$|e^{-kP_k(z)}| \leq e^{-k|z - z_0| + 1}, \qquad z \in X.$$

It follows that $f_k(z) \to -1/z - z_0$ for all $z \in X \setminus \{z_0\}$, as $k \to \infty$, and also

$$|f_k(z)| \leq \frac{4}{|z - z_0|}, \qquad z \in X.$$

Since by (15) $1/|z - z_0|$ is summable with respect to $|\mu|$, this implies that

$$\int f_k \, d\mu \to -\int \frac{d\mu(z)}{z - z_0}$$

by dominated convergence.

(18) then gives that

$$\int \frac{d\mu(z)}{z - z_0} = 0.$$

Since (15) holds a.e. on X by Lemma 2.4, and since certainly

$$\int \frac{d\mu(z)}{z - z_0} = 0 \qquad \text{for } z_0 \in \mathbf{C} \setminus X$$

(why?), we conclude that $\hat{\mu} = 0$ a.e., so $\mu = 0$ by Lemma 2.7. Thus $\mu \perp P(X) \Rightarrow \mu = 0$, and so $P(X) = C(X)$. Q.E.D.

NOTES

Proposition 2.2 is a part of the Krein–Milman theorem [4, p. 440]. The proof of Theorem 2.1 given here is due to de Branges [3]. Lemma 2.7 (concerning $\hat{\mu}$) is given by Bishop in [1]. Theorem 2.8 is in F. Hartogs and A. Rosenthal, Über Folgen analytischer Funktionen, *Math. Ann.* **104** (1931). Theorem 2.9 is due to C. Runge, Zur Theorie der eindeutigen analytischer Funktionen, *Acta Math.* **6** (1885). The proof given here is found in [8, Chap. 1]. Theorem 2.11 was proved by M. A. Lavrentieff, Sur les fonctions d'une variable complexe représentables par des séries de polynomes, Hermann, Paris, 1936, and a simpler proof is due to S. N. Mergelyan, On a theorem of M. A. Lavrentieff, *A.M.S. Transl.* **86** (1953). Lemma 2.12 and its use in the proof of Theorem 2.11 is in L. Carleson, Mergelyan's theorem on uniform polynomial approximation, *Math. Scand.* **15** (1964), 167–175.

Theorem 2.1 is due to M. H. Stone, Applications of the theory of Boolean rings to general topology, *Trans. Am. Math. Soc.* **41** (1937). See also M. H. Stone, The generalized Weierstrass approximation theorem, *Math. Mag.* **21** (1947–1948).

3

OPERATIONAL CALCULUS IN ONE VARIABLE

Let $\mathscr{F}$ denote the algebra of all functions f on $-\pi \leq \theta \leq \pi$, with

$$f(\theta) = \sum_{-\infty}^{\infty} C_n e^{in\theta}, \qquad \sum_{-\infty}^{\infty} |C_n| < \infty.$$

Exercise 3.1. $\mathscr{M}(\mathscr{F})$ may be identified with the circle $|\zeta| = 1$ and for $f = \Sigma_{-\infty}^{\infty} C_n e^{in\theta}$, $|\zeta_0| = 1$,

$$\hat{f}(\zeta_0) = \sum_{-\infty}^{\infty} C_n \zeta_0^n.$$

If $f \in \mathscr{F}$ and f never vanishes on $-\pi \leq \theta \leq \pi$, it follows that $\hat{f} \neq 0$ on $\mathscr{M}(\mathscr{F})$ and so that f has an inverse in $\mathscr{F}$, i.e.,

$$\frac{1}{f} = \sum_{-\infty}^{\infty} d_n e^{in\theta}$$

with $\Sigma_{-\infty}^{\infty} |d_n| < \infty$.

This result, that nonvanishing elements of $\mathscr{F}$ have inverses in $\mathscr{F}$, is due to Wiener (see [11, p. 91]), by a quite different method.

We now ask: Fix $f \in \mathscr{F}$ and let σ be the range of f; i.e., $\sigma = \{f(\theta) | -\pi \leq \theta \leq \pi\}$. Let Φ be a continuous function defined on σ, so that $\Phi(f)$ is a continuous function on $[-\pi, \pi]$. Does $\Phi(f) \in \mathscr{F}$?

The preceding result concerned the case $\Phi(z) = 1/z$.

Lévy [10] extended Wiener's result as follows: Assume that Φ is holomorphic in a neighborhood of σ. Then $\Phi(f) \in \mathscr{F}$.

How can we generalize this result to arbitrary Banach algebras?

THEOREM 3.1

Let $\mathfrak{A}$ be a Banach algebra and fix $x \in \mathfrak{A}$. Let $\sigma(x)$ denote the spectrum of x. If Φ is any function holomorphic in a neighborhood of $\sigma(x)$, then $\Phi(\hat{x}) \in \hat{\mathfrak{A}}$.

Note that this contains Lévy's theorem. However, we should like to do better. We want to define an element $\Phi(x) \in \mathfrak{A}$ so as to get a well-behaved map: $\Phi \to \Phi(x)$, not merely to consider the function $\Phi(\hat{x})$ on $\mathcal{M}$. When $\mathfrak{A}$ is not semisimple, this becomes important. We demand that

$$\widehat{\Phi(x)} = \Phi(\hat{x}) \text{ on } \mathcal{M}. \tag{1}$$

The study of a map $\Phi \to \Phi(x)$, from $H(\Omega) \to \mathfrak{A}$, we call the *operational calculus (in one variable)*.

For certain holomorphic functions Φ it is obvious how to define $\Phi(x)$. Let Φ be a polynomial

$$\Phi(z) = \sum_{n=0}^{N} a_n z^n.$$

We put

$$\Phi(x) = \sum_{n=0}^{N} a_n x^n. \tag{2}$$

Note that (1) holds. Let Φ be a rational function holomorphic on $\sigma(x)$,

$$\Phi(z) = \frac{P(z)}{Q(z)},$$

P and Q being polynomials and $Q(z) \neq 0$ for $z \in \sigma(x)$. Then

$$(Q(x))^{-1} \in \mathfrak{A} \qquad \text{(why?)}$$

and we define

$$\Phi(x) = P(x) \cdot Q(x)^{-1}. \tag{3}$$

We again verify (1).

Now let Ω be an open set with $\sigma(x) \subset \Omega$ and fix $\Phi \in H(\Omega)$. It follows from Theorem 2.9 that we can choose a sequence $\{f_n\}$ of rational functions holomorphic in Ω such that $f_n \to \Phi$ uniformly on compact subsets of Ω. (Why?) For each n, $f_n(x)$ was defined above. We want to define

$$\Phi(x) = \lim_{n \to \infty} f_n(x).$$

To do this, we must prove

LEMMA 3.2

$\lim_{n \to \infty} f_n(x)$ *exists in $\mathfrak{A}$ and depends only on x and Φ, not on the choice of $\{f_n\}$.*

We need

***Exercise 3.2.** Let $x \in \mathfrak{A}$, let Ω be an open set containing $\sigma(x)$, and let f be a rational functional holomorphic in Ω.

Choose an open set Ω_1 with

$$\sigma(x) \subset \Omega_1 \subset \bar{\Omega}_1 \subset \Omega$$

whose boundary γ is the union of finitely many simple closed polygonal curves. Then

$$f(x) = \frac{1}{2\pi i} \int_\gamma f(t) \cdot (t - x)^{-1} dt. \tag{4}$$

Proof of Lemma 3.2. Choose γ as in Exercise 3.2. Then

$$\begin{aligned} \left\| f_n(x) - \frac{1}{2\pi i} \int_\gamma \frac{\Phi(t)\, dt}{t - x} \right\| &= \left\| \frac{1}{2\pi i} \int_\gamma \frac{f_n(t) - \Phi(t)}{t - x} dt \right\| \\ &\leq \frac{1}{2\pi} \int_\gamma |f_n(t) - \Phi(t)| \, \|(t - x)^{-1}\| \, ds \end{aligned}$$

$\to 0$ as $n \to \infty$, since $\|(t - x)^{-1}\|$ is bounded on γ while $f_n \to \Phi$ uniformly on γ. Thus

$$\lim_{n \to \infty} f_n(x) = \frac{1}{2\pi i} \int_\gamma \frac{\Phi(t)\, dt}{t - x}. \tag{5}$$

Q.E.D.

Now let $\{F_n\}$ be a sequence in $H(\Omega)$. We write

$$F_n \to F \text{ in } H(\Omega)$$

if F_n tends to F uniformly on compact sets in Ω.

THEOREM 3.3

Let $\mathfrak{A}$ be a Banach algebra, $x \in \mathfrak{A}$, and let Ω be an open set containing $\sigma(x)$. Then there exists a map $\tau : H(\Omega) \to \mathfrak{A}$ such that the following holds. We write $F(x)$ for $\tau(F)$:

(a) *τ is an algebraic homomorphism.*
(b) *If $F_n \to F$ in $H(\Omega)$, then $F_n(x) \to F(x)$ in $\mathfrak{A}$.*
(c) *$\widehat{F(x)} = F(\hat{x})$ for all $F \in H(\Omega)$.*
(d) *If F is the identity function, $F(x) = x$.*
(e) *With γ as earlier, if $F \in H(\Omega)$,*

$$F(x) = \frac{1}{2\pi i} \int_\gamma \frac{F(t)\, dt}{t - x}.$$

Properties (a), (b), *and* (d) *define τ uniquely.*

Note. Theorem 3.1 is contained in this result.

Proof. Fix $F \in H(\Omega)$. Choose a sequence of rational functions $\{f_n\} \in H(\Omega)$ with $f_n \to F$ in $H(\Omega)$. By Lemma 3.2

$$\lim_{n\to\infty} f_n(x) \tag{6}$$

exists in $\mathfrak{A}$. We define this limit to be $F(x)$ and τ to be the map $F \to F(x)$.

τ is evidently a homomorphism when restricted to rational functions. Equation (6) then yields (a). Similarly, (c) holds for rational functions and so by (6) in general. Part (d) follows from (6).

Part (e) coincides with (5). Part (b) comes from (e) by direct computation.

Suppose now that τ' is a map from $H(\Omega)$ to $\mathfrak{A}$ satisfying (a), (b), and (d).

By (a) and (d), τ' and τ agree on rational functions. By (b), then, $\tau' = \tau$ on $H(\Omega)$.

Q.E.D.

We now consider some consequences of Theorem 3.3 as well as some related questions.

Let $\mathfrak{A}$ be a Banach algebra. By a *nontrivial idempotent e in* $\mathfrak{A}$ we mean an element e with $e^2 = e$, e not the zero element or the identity. Suppose that e is such an element. Then $1 - e$ is another. e is not in the radical (why?), so $\hat{e} \not\equiv 0$ on $\mathcal{M}$. Similarly, $\widehat{1 - e} \not\equiv 0$, so $\hat{e} \not\equiv 1$. But $\hat{e}^2 = \hat{e}$, so $\hat{e}$ takes on only the values 0 and 1 on $\mathcal{M}$. It follows that $\mathcal{M}$ is disconnected.

Question. Does the converse hold? That is, if $\mathcal{M}$ is disconnected, must $\mathfrak{A}$ contain a nontrivial idempotent?

At this moment, we can prove only a weaker result.

COROLLARY

Assume there is an element x in $\mathfrak{A}$ *such that* $\sigma(x)$ *is disconnected. Then* $\mathfrak{A}$ *contains a nontrivial idempotent.*

Proof. $\sigma(x) = K_1 \cup K_2$, where K_1, K_2 are disjoint closed sets. Choose disjoint open sets Ω_1 and Ω_2,

$$K_1 \subset \Omega_1, \qquad K_2 \subset \Omega_2.$$

Put $\Omega = \Omega_1 \cup \Omega_2$. Define F on Ω by

$$F = 1 \text{ on } \Omega_1, \qquad F = 0 \text{ on } \Omega_2.$$

Then $F \in H(\Omega)$. Put

$$e = F(x).$$

By Theorem 3.3,

$$e^2 = F^2(x) = F(x) = e$$

and

$$\hat{e} = F(\hat{x}) = \begin{cases} 1 & \text{on } \hat{x}^{-1}(K_1), \\ 0 & \text{on } \hat{x}^{-1}(K_2). \end{cases}$$

Hence e is a nontrivial idempotent.

Exercise 3.3. Let B be a Banach space and T a bounded linear operator on B having disconnected spectrum. Then there exists a bounded linear operator E on B, $E \neq 0$, $E \neq I$, such that $E^2 = E$ and E commutes with T.

Exercise 3.4. Let $\mathfrak{A}$ be a Banach algebra. Assume that $\mathscr{M}$ is a finite set. Then there exist idempotents $e_1, e_2, \ldots, e_n \in \mathfrak{A}$ with $e_i \cdot e_j = 0$ if $i \neq j$ and with $\Sigma_{i=1}^n e_i = 1$ such that the following holds:

Every x in $\mathfrak{A}$ admits a representation

$$x = \sum_{i=1}^{n} \lambda_i e_i + \rho,$$

where the λ_i are scalars and ρ is in the radical.

Note. Exercise 3.4 contains the following classical fact: If α is an $n \times n$ matrix with complex entries, then there exist commuting matrices β and γ with β nilpotent, γ diagonalizable, and

$$\alpha = \beta + \gamma.$$

To see this, put $\mathfrak{A}$ = algebra of all polynomials in α, normed so as to be a Banach algebra, and apply the exercise.

We consider another problem. Given a Banach algebra $\mathfrak{A}$ and an invertible element $x \in \mathfrak{A}$, when can we find $y \in \mathfrak{A}$ so that

$$x = e^y?$$

There is a purely topological necessary condition: There must exist f in $C(\mathscr{M})$ so that

$$\hat{x} = e^f \text{ on } \mathscr{M}.$$

(Think of an example where this condition is not satisfied.)

We can give a sufficient condition:

COROLLARY

Assume that $\sigma(x)$ is contained in a simply connected region Ω, where $0 \notin \Omega$. Then there is a y in $\mathfrak{A}$ with $x = e^y$.

Proof. Let Φ be a single-valued branch of $\log z$ defined in Ω. Put $y = \Phi(x)$.

$$\sum_0^N \frac{\Phi^n}{n!} \to e^\Phi = z \text{ in } H(\Omega), \qquad \text{as } N \to \infty.$$

Hence by Theorem 3.3(b),

$$\left(\sum_0^N \frac{\Phi^n}{n!}\right)(x) \to x.$$

By (a) the left side equals

$$\sum_0^N \frac{(\Phi(x))^n}{n!} \to e^y.$$

Hence $e^y = x$.

To find complete answers to the questions about existence of idempotents and representation of elements as exponentials, we need some more machinery.

We shall develop this machinery, concerning differential forms and the $\bar{\partial}$-operator, in the next three sections. We shall then use the machinery to set up an operational calculus in several variables for Banach algebras, to answer the above questions, and to attack various other problems.

NOTES

Theorem 3.3 has a long history. See E. Hille and R. S. Phillips, Functional analysis and semi-groups, *Am. Math. Soc. Coll. Publ. XXXI*, 1957, Chap. V. In the form given here, it is part of Gelfand's theory [5]. For the result on idempotents and related results, see Hille and Phillips, *loc. cit.*

4

DIFFERENTIAL FORMS

Note. The proofs of all lemmas in this section are left as exercises.

The notion of differential form is defined for arbitrary differentiable manifolds. For our purposes, it will suffice to study differential forms on an open subset Ω of real Euclidean N-space R^N. Fix such an Ω. Denote by $x_1, \ldots, x_N$ the coordinates in R^N.

Definition 4.1. $C^\infty(\Omega)$ = algebra of all infinitely differentiable complex-valued functions on Ω.

We write C^∞ for $C^\infty(\Omega)$.

Definition 4.2. Fix $x \in \Omega$. T_x is the collection of all maps $v : C^\infty \to \mathsf{C}$ for which

(a) v is linear.

(b) $v(f \cdot g) = f(x) \cdot v(g) + g(x) \cdot v(f)$, $f, g \in C^\infty$.

T_x evidently forms a vector space over C. We call it the *tangent space* at x and its elements *tangent vectors* at x.

Denote by $\partial/\partial x_j|_x$ the functional $f \to \dfrac{\partial f}{\partial x_j}(x)$. Then $\partial/\partial x_j|_x$ is a tangent vector at x for $j = 1, 2, \ldots, n$.

LEMMA 4.1

$\partial/\partial x_1|_x, \ldots, \partial/\partial x_N|_x$ forms a basis for T_x.

Definition 4.3. The dual space to T_x is denoted T_x^*.

Note. The dimension of T_x^* over C is N.

Definition 4.4. A 1-*form* ω on Ω is a map ω assigning to each x in Ω an element of T_x^*.

Example. Let $f \in C^\infty$. For $x \in \Omega$, put

$$(df)_x(v) = v(f), \qquad \text{all } v \in T_x.$$

Then $(df)_x \in T_x^*$.

df is the 1-form on Ω assigning to each x in Ω the element $(df)_x$.

Note. $dx_1, \ldots, dx_N$ are particular 1-forms. In a natural way 1-forms may be added and multiplied by scalar functions.

LEMMA 4.2

Every 1-form ω admits a unique representation

$$\omega = \sum_1^N C_j \, dx_j,$$

the C_j being scalar functions on Ω.

Note. For $f \in C^\infty$,

$$df = \sum_{j=1}^N \frac{\partial f}{dx_j} dx_j.$$

We now recall some multilinear algebra. Let V be an N-dimensional vector space over $\mathbf{C}$. Denote by $\Lambda^k(V)$ the vector space of k-linear alternating maps of $V \times \cdots \times V \to \mathbf{C}$. ("Alternating" means that the value of the function changes sign if two of the variables are interchanged.)

Define $\mathscr{G}(V)$ as the direct sum

$$\mathscr{G}(V) = \Lambda^0(V) \oplus \Lambda^1(V) \oplus \cdots \oplus \Lambda^N(V).$$

Here $\Lambda^0(V) = \mathbf{C}$ and $\Lambda^1(V)$ is the dual space of V. Put $\Lambda^j(V) = 0$ for $j > N$.

We now introduce a multiplication into the vector space $\mathscr{G}(V)$. Fix $\tau \in \Lambda^k(V)$, $\sigma \in \Lambda^l(V)$. The map

$$(\xi_1, \ldots, \xi_k, \xi_{k+1}, \ldots, \xi_{k+l}) \to \tau(\xi_1, \ldots, \xi_k)\sigma(\xi_{k+1}, \ldots, \xi_{k+l})$$

is a $(k + l)$-linear map from $V \times \cdots \times V$ ($k + l$ factors) $\to \mathbf{C}$. It is, however, not alternating. To obtain an alternating map, we use

Definition 4.5. Let $\tau \in \Lambda^k(V)$, $\sigma \in \Lambda^l(V)$, $k,l \geq 1$.

$$\tau \wedge \sigma(\xi_1, \ldots, \xi_{k+l}) = \frac{1}{(k+l)!} \sum_\pi (-1)^\pi \tau(\xi_{\pi(1)}, \ldots, \xi_{\pi(k)}) \cdot \sigma(\xi_{\pi(k+1)}, \ldots, \xi_{\pi(k+l)}),$$

the sum being taken over all permutations π of the set $\{1, 2, \ldots, k + l\}$, and $(-1)^\pi$ denoting the sign of the permutation π.

LEMMA 4.3

$\tau \wedge \sigma$ as defined is $(k + l)$-linear and alternating and so $\in \Lambda^{k+l}(V)$.

The operation $\wedge$ (wedge) defines a product for pairs of elements, one in $\Lambda^k(V)$ and one in $\Lambda^l(V)$, the value lying in $\Lambda^{k+l}(V)$, hence in $\mathscr{G}(V)$. By linearity, $\wedge$ extends to a product on arbitrary pairs of elements of $\mathscr{G}(V)$ with value in $\mathscr{G}(V)$. For $\tau \in \Lambda^0(V)$, $\sigma \in \mathscr{G}(V)$, define $\tau \wedge \sigma$ as scalar multiplication by τ.

LEMMA 4.4

Under $\wedge$, $\mathscr{G}(V)$ is an associative algebra with identity.

$\mathscr{G}(V)$ is not commutative. In fact,

LEMMA 4.5

If $\tau \in \Lambda^k(V)$, $\sigma \in \Lambda^l(V)$, then $\tau \wedge \sigma = (-1)^{kl}\sigma \wedge \tau$.

Let $e_1, \ldots, e_N$ form a basis for $\Lambda^1(V)$.

LEMMA 4.6

Fix k. The set of elements

$$e_{i_1} \wedge e_{i_2} \wedge \cdots \wedge e_{i_k}, \qquad 1 \le i_1 < i_2 < \cdots < i_k \le N,$$

forms a basis for $\Lambda^k(V)$.

We now apply the preceding to the case when $V = T_x$, $x \in \Omega$. Then $\Lambda^k(T_x)$ is the space of all k-linear alternating functions on T_x, and so, for $k = 1$, coincides with T_x^*. The following thus extends our definition of a 1-form.

Definition 4.6. A *k-form* ω^k on Ω is a map ω^k assigning to each x in Ω an element of $\Lambda^k(T_x)$.

k-forms form a module over the algebra of scalar functions on Ω in a natural way.

Let τ^k and σ^l be, respectively, a k-form and an l-form. For $x \in \Omega$, put

$$\tau^k \wedge \sigma^l(x) = \tau^k(x) \wedge \sigma^l(x) \in \Lambda^{k+l}(T_x).$$

In particular, since $dx_1, \ldots, dx_N$ are 1-forms,

$$dx_{i_1} \wedge dx_{i_2} \wedge \cdots \wedge dx_{i_k}$$

is a k-form for each choice of $(i_1, \ldots, i_k)$.

Because of Lemma 4.5,

$$dx_j \wedge dx_j = 0 \text{ for each } j.$$

Hence $dx_{i_1} \wedge \cdots \wedge dx_{i_k} = 0$ unless the i_v are distinct.

LEMMA 4.7

Let ω^k be any k-form on Ω. Then there exist (unique) scalar functions $C_{i_1}, \ldots, i_k$ on Ω such that

$$\omega^k = \sum_{i_1 < i_2 < \cdots < i_k} C_{i_1} \cdots i_k \, dx_{i_1} \wedge \cdots \wedge dx_{i_k}.$$

Definition 4.7. $\Lambda^k(\Omega)$ consists of all k-forms ω^k such that the functions $C_{i_1} \cdots i_k$ occurring in Lemma 4.7 lie in C^∞. $\Lambda^0(\Omega) = C^\infty$.

Recall now the map $f \to df$ from $C^\infty \to \Lambda^1(\Omega)$. We wish to extend d to a linear map $\Lambda^k(\Omega) \to \Lambda^{k+1}(\Omega)$, for all k.

Definition 4.8. Let $\omega^k \in \Lambda^k(\Omega)$, $k = 0, 1, 2, \ldots$. Then

$$\omega^k = \sum_{i_1 < \cdots < i_k} C_{i_1} \cdots {}_{i_k}\, dx_{i_1} \wedge \cdots \wedge dx_{i_k}.$$

Define

$$d\omega^k = \sum_{i_1 < \cdots < i_k} dC_{i_1} \cdots {}_{i_k} \wedge dx_{i_1} \wedge \cdots \wedge dx_{i_k}.$$

Note that d maps $\Lambda^k(\Omega) \to \Lambda^{k+1}(\Omega)$. We call $d\omega^k$ the *exterior derivative* of ω^k. For $\omega \in \Lambda^1(\Omega)$,

$$\omega = \sum_{i=1}^{N} C_i\, dx_i,$$

$$d\omega = \sum_{i,j} \frac{\partial C_i}{\partial x_j} dx_j \wedge dx_i = \sum_{i<j} \left(\frac{\partial C_j}{\partial x_i} - \frac{\partial C_i}{\partial x_j} \right) dx_i \wedge dx_j.$$

It follows that for $f \in C^\infty$,

$$d(df) = d\left(\sum_{i=1}^{N} \frac{\partial f}{dx_i} dx_i \right) = \sum_{i<j} \left(\frac{\partial}{\partial x_i}\left(\frac{\partial f}{\partial x_j}\right) - \frac{\partial}{\partial x_j}\left(\frac{\partial f}{\partial x_i}\right) \right) dx_i \wedge dx_j = 0$$

or $d^2 = 0$ on C^∞. More generally,

LEMMA 4.8

$d^2 = 0$ *for every* k; i.e., *if* $\omega^k \in \Lambda^k(\Omega)$, *$k$ arbitrary, then* $d(d\omega^k) = 0$.

To prove Lemma 4.8, it is useful to prove first

LEMMA 4.9

Let $\omega^k \in \Lambda^k(\Omega)$, $\omega^l \in \Lambda^l(\omega)$. *Then*

$$d(\omega^k \wedge \omega^l) = d\omega^k \wedge \omega^l + (-1)^k \omega^k \wedge d\omega^l.$$

NOTES

For an exposition of the material in this section, see, e.g., I. M. Singer and J. A. Thorpe, *Lecture Notes on Elementary Topology and Geometry*, Scott, Foresman, Glenview, Ill., 1967, Chap. V.

5

THE $\bar{\partial}$-OPERATOR

Note. As in the preceding section, the proofs in this section are left as exercises.

Let Ω be an open subset of $\mathbf{C}^n$.

The complex coordinate functions $z_1, \ldots, z_n$ as well as their conjugates $\bar{z}_1, \ldots, \bar{z}_n$ lie in $C^\infty(\Omega)$. Hence the forms

$$dz_1, \ldots, dz_n, \qquad d\bar{z}_1, \ldots, d\bar{z}_n$$

all belong to $\Lambda^1(\Omega)$. Fix $x \in \Omega$. Note that $\Lambda^1(T_x) = T_x^*$ has dimension $2n$ over $\mathbf{C}$, since $\mathbf{C}^n = \mathbf{R}^{2n}$. If $x_j = \operatorname{Re}(z_j)$ and $y_j = \operatorname{Im}(z_j)$, then

$$(dx_1)_x, \ldots, (dx_n)_x, \qquad (dy_1)_x, \ldots, (dy_n)_x$$

form a basis for T_x^*. Since $dx_j = 1/2(dz_j + d\bar{z}_j)$ and $dy_j = 1/2i(dz_j - d\bar{z}_j)$,

$$(dz_1)_x, \ldots, (dz_n)_x, \qquad (d\bar{z}_1)_x, \ldots, (d\bar{z}_n)_x$$

also form a basis for T_x^*. In fact,

LEMMA 5.1

If $\omega \in \Lambda^1(\Omega)$, then

$$\omega = \sum_{j=1}^{n} a_j\, dz_j + b_j\, d\bar{z}_j,$$

where $a_j, b_j \in C^\infty$.

Fix $f \in C^\infty$. Since $(x_1, \ldots, x_n, y_1, \ldots, y_n)$ are real coordinates in $\mathbf{C}^n$,

$$\begin{aligned} df &= \sum_{j=1}^{n} \frac{\partial f}{\partial x_j} dx_j + \frac{\partial f}{\partial y_j} dy_j \\ &= \sum_{j=1}^{n} \left(\frac{\partial f}{\partial x_j} \cdot \frac{1}{2} + \frac{\partial f}{\partial y_j} \cdot \frac{1}{2i} \right) dz_j + \left(\frac{\partial f}{\partial x_j} \cdot \frac{1}{2} - \frac{1}{2i} \frac{\partial f}{\partial y_j} \right) d\bar{z}_j. \end{aligned}$$

Definition 5.1. We define operators on C^∞ as follows:

$$\frac{\partial}{dz_j} = \frac{1}{2}\left(\frac{\partial}{\partial x_j} - i\frac{\partial}{\partial y_j}\right), \qquad \frac{\partial}{\partial \bar{z}_j} = \frac{1}{2}\left(\frac{\partial}{\partial x_j} + i\frac{\partial}{\partial y_j}\right).$$

Then for $f \in C^\infty$,

$$df = \sum_{j=1}^{n} \frac{\partial f}{\partial z_j} dz_j + \frac{\partial f}{\partial \bar{z}_j} d\bar{z}_j. \tag{1}$$

Definition 5.2. We define two maps from $C^\infty \to \Lambda^1(\Omega)$, ∂ and $\bar{\partial}$. For $f \in C^\infty$,

$$\partial f = \sum_{j=1}^{n} \frac{\partial f}{\partial z_j} dz_j, \qquad \bar{\partial} f = \sum_{j=1}^{n} \frac{\partial f}{\partial \bar{z}_j} d\bar{z}_j.$$

Note. $\partial f + \bar{\partial} f = df$, if $f \in C^\infty$.

We need some notation. Let I be any r-tuple of integers, $I = (i_1, i_2, \ldots, i_r)$, $1 \leq i_j \leq n$, all j. Put

$$dz_I = dz_{i_1} \wedge \cdots \wedge dz_{i_r}.$$

Thus $dz_I \in \Lambda^r(\Omega)$.

Let J be any s-tuple $(j_1, \ldots, j_s)$, $1 \leq j_k \leq n$, all k, and put

$$d\bar{z}_J = d\bar{z}_{j_1} \wedge \cdots \wedge d\bar{z}_{j_s}.$$

So $d\bar{z}_J \in \Lambda^s(\Omega)$. Then

$$dz_I \wedge d\bar{z}_J \in \Lambda^{r+s}(\Omega).$$

For I as above, put $|I| = r$. Then $|J| = s$.

Definition 5.3. Fix integers $r,s \geq 0$. $\Lambda^{r,s}(\Omega)$ is the space of all $\omega \in \Lambda^{r+s}(\Omega)$ such that

$$\omega = \sum_{I,J} a_{IJ}\, dz_I \wedge d\bar{z}_J,$$

the sum being extended over all I,J with $|I| = r$, $|J| = s$, and with each $a_{IJ} \in C^\infty$.

An element of $\Lambda^{r,s}(\Omega)$ is called a *form of type* (r,s). We now have a direct sum decomposition of each $\Lambda^k(\Omega)$:

LEMMA 5.2

$$\Lambda^k(\Omega) = \Lambda^{0,k}(\Omega) \oplus \Lambda^{1,k-1}(\Omega) \oplus \Lambda^{2,k-2}(\Omega) \oplus \cdots \oplus \Lambda^{k,0}(\Omega).$$

We extend the definition of ∂ and $\bar{\partial}$ (see Definition 5.2) to maps from $\Lambda^k(\Omega) \to \Lambda^{k+1}(\Omega)$ for all k, as follows:

Definition 5.4. Choose ω^k in $\Lambda^k(\omega)$,

$$\omega^k = \sum_{I,J} a_{IJ}\, dz_I \wedge d\bar{z}_J,$$

$$\partial\omega^k = \sum_{I,J} \partial a_{IJ} \wedge dz_I \wedge d\bar{z}_J,$$

and

$$\bar{\partial}\omega^k = \sum_{I,J} \bar{\partial}a_{IJ} \wedge dz_I \wedge d\bar{z}_J.$$

Observe that, by (1), if ω^k is as above,

$$\bar{\partial}\omega^k + \partial\omega^k = \sum_{I,J} da_{IJ} \wedge dz_I \wedge d\bar{z}_J = d\omega^k,$$

so we have

(2) $$\bar{\partial} + \partial = d$$

as operators from $\Lambda^k(\Omega) \to \Lambda^{k+1}(\Omega)$. Note that if $\omega \in \Lambda^{r,s}$, $\partial\omega \in \Lambda^{r+1,s}$ and $\bar{\partial}\omega \in \Lambda^{r,s+1}$.

LEMMA 5.3

$\bar{\partial}^2 = 0$, $\partial^2 = 0$, *and* $\partial\bar{\partial} = \bar{\partial}\partial = 0$.

Why is the $\bar{\partial}$-operator of interest to us? Consider $\bar{\partial}$ as the map from $C^\infty \to \Lambda^1(\Omega)$. What is its kernel?

Let $f \in C^\infty$. $\bar{\partial}f = 0$ if and only if

(3) $$\frac{\partial f}{\partial \bar{z}_j} = 0 \text{ in } \Omega, \qquad j = 1, 2, \ldots, n.$$

For $n = 1$ and Ω a domain in the z-plane, (3) reduces to

$$\frac{df}{\partial \bar{z}} = 0 \qquad \text{or } \frac{\partial f}{\partial x} + i\frac{\partial f}{\partial y} = 0.$$

For $f = u + iv$, u and v real-valued, this means that

$$\frac{\partial u}{\partial x} = \frac{\partial v}{\partial y}, \qquad \frac{\partial v}{\partial x} = -\frac{\partial u}{\partial y},$$

or u and v satisfy the Cauchy–Riemann equations. Thus here

$$\bar{\partial}f = 0 \text{ in } \Omega \text{ is equivalent to } f \in H(\Omega).$$

Definition 5.5. Let Ω be an open subset of $\mathbf{C}^n$. $H(\Omega)$ is the class of all $f \in C^\infty$ with $\bar{\partial}f = 0$ in Ω, or, equivalently, (3).

We call the elements of $H(\Omega)$ *holomorphic* in Ω. Note that, by (3), $f \in H(\Omega)$ if and only if f is holomorphic in each fixed variable z_j (as the function of a single complex variable), when the remaining variables are held fixed.

Let now Ω be the domain

$$\{z \in \mathbf{C}^n \mid |z_j| < R_j, j = 1, \ldots, n\},$$

where $R_1, \ldots, R_n$ are given positive numbers. Thus Ω is a product of n open plane disks. Let f be a once-differentiable function on Ω; i.e., $df/\partial x_j$ and $df/\partial y_j$ exist and are continuous in Ω, $j = 1, \ldots, n$.

LEMMA 5.4

Assume that $\partial f/\partial \bar{z}_j = 0$, $j = 1, \ldots, n$, *in* Ω. *Then there exist constants* A_ν *in* $\mathbf{C}$ *for each tuple* $\nu = (\nu_1, \ldots, \nu_n)$ *of nonnegative integers such that*

$$f(z) = \sum_\nu A_\nu z^\nu,$$

where $z^\nu = z_1^{\nu_1} \cdot z_2^{\nu_2} \cdots z_n^{\nu_n}$, *the series converging absolutely in* Ω *and uniformly on every compact subset of* Ω.

For a proof of this result, see, e.g., [8, Th. 2.2.6].

This result then applies in particular to every f in $H(\Omega)$. We call $\Sigma_\nu A_\nu z^\nu$ the *Taylor series* for f at 0.

We shall see that the study of the $\bar{\partial}$-operator, to be undertaken in the next section and in later sections, will throw light on the holomorphic functions of several complex variables.

For future use, note also

LEMMA 5.5

If $\omega^k \in \Lambda^k(\Omega)$ *and* $\omega^l \in \Lambda^l(\Omega)$, *then*

$$\bar{\partial}(\omega^k \wedge \omega^l) = \bar{\partial}\omega^k \wedge \omega^l + (-1)^k \omega^k \wedge \bar{\partial}\omega^l.$$

6

THE EQUATION $\bar{\partial} u = f$

As before, fix an open set $\Omega \subset \mathbf{C}^n$. Given $f \in \Lambda^{r,s+1}(\Omega)$, we seek $u \in \Lambda^{r,s}$ such that

$$\bar{\partial} u = f. \tag{1}$$

Since $\bar{\partial}^2 = 0$ (Lemma 5.3), a necessary condition on f is

$$\bar{\partial} f = 0. \tag{2}$$

If (2) holds, we say that f is *$\bar{\partial}$-closed*. What is a sufficient condition on f? It turns out that this will depend on the domain Ω.

Recall the analogous problem for the operator d on a domain $\Omega \subset \mathbf{R}^n$. If ω^k is a k-form in $\Lambda^k(\Omega)$, the condition

$$d\omega^k = 0 \qquad (\omega \text{ is “closed”}) \tag{3}$$

is necessary in order that we can find some τ^{k-1} in $\Lambda^{k-1}(\Omega)$ with

$$d\tau^{k-1} = \omega^k. \tag{4}$$

However, (3) is, in general, not sufficient. (Think of an example when $k = 1$ and Ω is an annulus in $\mathbf{R}^2$.) If Ω is contractible, then (3) is sufficient in order that (4) admit a solution.

For the $\bar{\partial}$-operator, a purely topological condition on Ω is inadequate. We shall find various conditions in order that (1) will have a solution. Denote by Δ^n the closed unit polydisk in $\mathbf{C}^n : \Delta^n = \{z \in \mathbf{C}^n \,|\, |z_j| \leq 1, j = 1, \ldots, n\}$.

THEOREM 6.1 (COMPLEX POINCARÉ LEMMA)

Let Ω be a neighborhood of Δ^n. Fix $\omega \in \Lambda^{p,q}(\Omega)$, $q > 0$, with $\bar{\partial}\omega = 0$. Then there exists a neighborhood Ω^* of Δ^n and there exists $\omega^* \in \Lambda^{p,q-1}(\Omega^*)$ such that

$$\bar{\partial}\omega^* = \omega \text{ in } \Omega^*.$$

We need some preliminary work.

LEMMA 6.2

Let $\phi \in C^1(\mathsf{R}^2)$ and assume that ϕ has compact support. Put

$$\Phi(\zeta) = -\frac{1}{\pi}\int_{\mathsf{R}^2} \phi(z)\frac{dx\,dy}{z-\zeta}.$$

Then $\Phi \in C^1(\mathsf{R}^2)$ and $\partial\Phi/\partial\bar{\zeta} = \phi(\zeta)$, all ζ.

Proof. Choose R with supp $\phi \subset \{z \mid |z| \le R\}$.

$$\pi\Phi(\zeta) = \int_{|z|\le R} \phi(z)\frac{1}{\zeta - z}\,dx\,dy = \int_{|z'-\zeta|\le R} \phi(\zeta - z')\frac{dx'\,dy'}{z'}$$
$$= \int_{\mathsf{R}^2} \phi(\zeta - z')\frac{dx'\,dy'}{z'}.$$

Since $1/z' \in L^1(dx'\,dy')$ on compact sets, it is legal to differentiate the last integral under the integral sign. We get

$$\pi\frac{\partial\Phi}{\partial\bar{\zeta}}(\zeta) = \int_{\mathsf{R}^2} \frac{\partial}{\partial\bar{\zeta}}[\phi(\zeta - z')]\frac{dx'\,dy'}{z'} = \int_{\mathsf{R}^2} \frac{\partial\phi}{\partial\bar{z}}(\zeta - z')\frac{dx'\,dy'}{z'}$$
$$= \int_{\mathsf{R}^2} \frac{\partial\phi}{\partial\bar{z}}(z)\frac{dx\,dy}{\zeta - z}.$$

On the other hand, Lemma 2.5 gives that

$$-\pi\phi(\zeta) = \int_{\mathsf{R}^2} \frac{\partial\phi}{\partial\bar{z}}(z)\frac{dx\,dy}{z-\zeta}.$$

Hence $\partial\Phi/\partial\bar{\zeta} = \phi$. Q.E.D.

LEMMA 6.3

Let Ω be a neighborhood of Δ^n and fix f in $C^\infty(\Omega)$. Fix j, $1 \le j \le n$. Assume that

$$\frac{\partial f}{\partial\bar{z}_k} = 0 \text{ in } \Omega,\ k = k_1, \ldots, k_s, \text{ each } k_i \ne j. \tag{5}$$

Then we can find a neighborhood Ω_1 of Δ^n and F in $C^\infty(\Omega_1)$ such that

(a) $\partial F/\partial\bar{\zeta}_j = f$ *in* Ω_1.
(b) $\partial F/\partial\bar{\zeta}_k = 0$ *in* Ω_1, $k = k_1, \ldots, k_s$.

Proof. Choose $\varepsilon > 0$ so that if $z = (z_1, \ldots, z_n) \in \mathbf{C}^n$ and $|z_\nu| < 1 + 2\varepsilon$ for all ν, then $z \in \Omega$.

Choose $\psi \in C^\infty(\mathbf{R}^2)$, having support contained in $\{z \mid |z| < 1 + 2\varepsilon\}$, with $\psi(z) = 1$ for $|z| < 1 + \varepsilon$. Put

$$F(\zeta_1, \ldots, \zeta_j, \ldots, \zeta_n) = -\frac{1}{\pi}\int_{\mathbf{R}^2} \psi(z) f(\zeta_1, \ldots, \zeta_{j-1}, z, \zeta_{j+1}, \ldots, \zeta_n)\frac{dx\,dy}{z - \zeta_j}.$$

For fixed $\zeta_1, \ldots, \zeta_{j-1}, \zeta_{j+1}, \ldots, \zeta_n$ with $|\zeta_\nu| < 1 + \varepsilon$, all ν, we now apply Lemma 6.2 with

$$\begin{aligned}\phi(z) &= \psi(z) f(\zeta_1, \ldots, \zeta_{j-1}, z, \zeta_{j+1}, \ldots, \zeta_n), \ |z| < 1 + 2\varepsilon \\ &= 0 \quad \text{outside supp } \psi.\end{aligned}$$

We obtain

$$\frac{\partial F}{\partial \bar{\zeta}_j}(\zeta_1, \ldots, \zeta_j, \ldots, \zeta_n) = \phi(\zeta_j) = f(\zeta_1, \ldots, \zeta_{j-1}, \zeta_j, \zeta_{j+1}, \ldots, \zeta_n),$$

if $|\zeta_j| < 1 + \varepsilon$, and so (a) holds with

$$\Omega_1 = \{\zeta \in \mathbf{C}^n \mid |\zeta_\nu| < 1 + \varepsilon, \text{ all } \nu\}.$$

Part (b) now follows directly from (5) by differentiation under the integral sign.

Proof of Theorem 6.1. We call a form

$$\sum_{I,J} C_{IJ}\, dz_I \wedge d\bar{z}_J$$

of *level* ν, if for some I and J with $J = (j_1, j_2, \ldots, \nu)$, where $j_1 < j_2 < \cdots < \nu$, we have $C_{IJ} \neq 0$; while for each I and J with $J = (j_1, \ldots, j_s)$, where $j_1 < \cdots < j_s$ and $j_s > \nu$, we have $C_{IJ} = 0$.

Consider first a form ω of level 1 such that $\bar{\partial}\omega = 0$. Then $\omega \in \Lambda^{p,1}(\Omega)$ for some p and we have

$$\omega = \sum_I a_I\, d\bar{z}_1 \wedge dz_I, \qquad a_I \in C^\infty(\Omega) \qquad \text{for each } I.$$

$$0 = \bar{\partial}\omega = \sum_{I,k} \frac{\partial a_I}{\partial \bar{z}_k} d\bar{z}_k \wedge d\bar{z}_1 \wedge dz_I.$$

Hence $(\partial a_I/\partial \bar{z}_k)\, d\bar{z}_k \wedge d\bar{z}_1 \wedge dz_I = 0$ for each k and I. It follows that

$$\frac{\partial a_I}{\partial \bar{z}_k} = 0, \qquad k \geq 2, \text{ all } I.$$

By Lemma 6.3 there exists for every I, A_I in $C^\infty(\Omega_1)$, Ω_1 being some neighborhood of Δ^n, such that

$$\frac{\partial A_I}{\partial \bar{z}_1} = a_I \quad \text{and} \quad \frac{\partial A_I}{\partial \bar{z}_k} = 0, \qquad k = 2, \ldots, n.$$

Put $\tilde{\omega} = \Sigma_I A_I\, dz_I \in \Lambda^{p,0}(\Omega_1)$.

$$\bar{\partial}\tilde{\omega} = \sum_{I,k} \frac{\partial A_I}{\partial \bar{z}_k} d\bar{z}_k \wedge dz_I = \omega.$$

We proceed by induction. Assume that the assertion of the theorem holds whenever ω is of level $\leq \nu - 1$ and consider ω of level ν. By hypothesis $\omega \in \Lambda^{p,q}(\Omega)$ and $\bar{\partial}\omega = 0$. We can find forms α and β of level $\leq \nu - 1$ so that

$$\omega = d\bar{z}_\nu \wedge \alpha + \beta \qquad \text{(why?)}.$$

$$0 = \bar{\partial}\omega = -d\bar{z}_\nu \wedge \bar{\partial}\alpha + \bar{\partial}\beta,$$

where we have used Lemma 5.5. So

$$0 = d\bar{z}_\nu \wedge \bar{\partial}\alpha - \bar{\partial}\beta. \tag{6}$$

Put

$$\alpha = \sum_{I,J} a_{IJ}\, dz_I \wedge d\bar{z}_J, \qquad \beta = \sum_{I,J} b_{IJ}\, dz_I \wedge d\bar{z}_J.$$

Equation (6) gives

$$\begin{aligned} 0 = d\bar{z}_\nu \wedge &\sum_{I,J,k} \frac{\partial a_{IJ}}{\partial \bar{z}_k} d\bar{z}_k \wedge dz_I \wedge d\bar{z}_J \\ &- \sum_{I,J,k} \frac{\partial b_{IJ}}{\partial \bar{z}_k} d\bar{z}_k \wedge dz_I \wedge d\bar{z}_J. \end{aligned} \tag{7}$$

Fix $k > \nu$, and look at the terms on the right side of (7) containing $d\bar{z}_\nu \wedge d\bar{z}_k$. Because α and β are of level $\leq \nu - 1$, these are the terms:

$$d\bar{z}_\nu \wedge \frac{\partial a_{IJ}}{\partial \bar{z}_k} d\bar{z}_k \wedge dz_I \wedge d\bar{z}_J.$$

It follows that for each I and J,

$$\frac{\partial a_{IJ}}{\partial \bar{z}_k} = 0, \qquad k > \nu.$$

By Lemma 6.3 there exists a neighborhood Ω_1 of Δ^n and, for each I and J, $A_{IJ} \in C^\infty(\Omega_1)$ with

$$\frac{\partial A_{IJ}}{\partial \bar{z}_\nu} = a_{IJ}, \qquad \frac{\partial A_{IJ}}{\partial \bar{z}_k} = 0, \qquad k > \nu.$$

Put

$$\omega_1 = \sum_{I,J} A_{IJ}\, dz_I \wedge d\bar{z}_J \in \Lambda^{p,q-1}(\Omega_1),$$

$$\begin{aligned} \bar{\partial}\omega_1 &= \sum_{I,J,k} \frac{\partial A_{IJ}}{\partial \bar{z}_k} d\bar{z}_k \wedge dz_I \wedge d\bar{z}_J \\ &= \sum_{I,J} a_{IJ}\, d\bar{z}_\nu \wedge dz_I \wedge d\bar{z}_J + \gamma, \end{aligned}$$

where γ is a form of level $\leq \nu - 1$. Thus

$$\bar{\partial}\omega_1 = d\bar{z}_\nu \wedge \alpha + \gamma.$$

Hence

$$\bar{\partial}\omega_1 - \omega = \gamma - \beta$$

is a form of level $\leq \nu - 1$. Also

$$\bar{\partial}(\gamma - \beta) = \bar{\partial}(\bar{\partial}\omega_1 - \omega) = 0.$$

By induction hypothesis, we can choose a neighborhood Ω_2 of Δ^n and $\tau \in \Lambda^{p,q-1}(\Omega_2)$ with $\bar{\partial}\tau = \gamma - \beta$. Then

$$\bar{\partial}(\omega_1 - \tau) = \bar{\partial}\omega_1 - \bar{\partial}\tau = \omega + (\gamma - \beta) - (\gamma - \beta) = \omega.$$

$\omega_1 - \tau$ is now the desired ω^*. Q.E.D.

NOTES

Theorem 6.1 is in P. Dolbeaut, Formes différentielles et cohomologie sur une variété analytique complexe, I, *Ann. Math.* **64** (1956), 83–130; II, *Ann. Math.* **65** (1957), 282–330. For the proof cf. [8, Chap. 2].

7

THE OKA–WEIL THEOREM

Let K be a compact set in the z-plane and denote by $P(K)$ the uniform closure on K of the polynomials in z.

THEOREM 7.1

Assume that $\mathbf{C} \setminus K$ *is connected. Let* F *be holomorphic in some neighborhood* Ω *of* K. *Then* $F|_K$ *is in* $P(K)$.

Proof. Let $\mathscr{L}$ denote the space of all finite linear combinations of functions $1/(z - a)^p$, where $a \in \mathbf{C} \setminus \Omega$, p an integer ≥ 0. By Runge's theorem (Theorem 2.9), $F|_K$ lies in the uniform closure of $\mathscr{L}$ on K. We claim that $\mathscr{L} \subset P(K)$. For let μ be a measure on K, $\mu \perp P(K)$. Then for $|a|$ large,

$$\int \frac{d\mu(z)}{z - a} = -\int \left(\sum_0^\infty \frac{z^n}{a^{n+1}} \right) d\mu = 0.$$

But the integral on the left is analytic as a function of a in $\mathbf{C} \setminus K$ and, since $\mathbf{C} \setminus K$ is connected, vanishes for all a in $\mathbf{C} \setminus K$. By differentiation,

$$\int \frac{d\mu(z)}{(z - a)^p} = 0, \qquad p = 1,2,\ldots, a \in \mathbf{C} \setminus K.$$

Thus $\mu \perp \mathscr{L}$, so $\mathscr{L} \subset P(K)$, as claimed. The theorem follows.

How can we generalize this result to the case when K is a compact subset of $\mathbf{C}^n$, $n > 1$?

What condition on K will assure the possibility of approximating arbitrary functions holomorphic in a neighborhood of K uniformly on K by polynomials in $z_1, \ldots, z_n$?

Note that the condition "$\mathbf{C}\setminus K$ is connected" is a purely topological restriction on K. No such purely topological restriction can suffice when $n > 1$. As an example, consider the two sets in $\mathbf{C}^2$.

$$K_1 = \{(e^{i\theta}, 0) | 0 \leq \theta \leq 2\pi\},$$
$$K_2 = \{(e^{i\theta}, e^{-i\theta}) | 0 \leq \theta \leq 2\pi\}.$$

The two sets are, topologically, circles. The function $F(z_1, z_2) = 1/z_1$ is holomorphic in a neighborhood of K_1.

Yet we cannot approximate F uniformly on K_1 by polynomials in z_1, z_2. (Why?) On the other hand, every continuous function on K_2 is uniformly approximable by polynomials in z_1, z_2. (Why?)

To obtain a general condition valid in $\mathbf{C}^n$ for all n, we rephrase the statement "$\mathbf{C}\setminus K$ is connected" as follows:

LEMMA 7.2

Let K be a compact set in $\mathbf{C}$. $\mathbf{C}\setminus K$ is connected if and only if for each $x^0 \in \mathbf{C}\setminus K$ we can find a polynomial P such that

$$|P(x^0)| > \max_K |P|. \tag{1}$$

Proof. If $\mathbf{C}\setminus K$ fails to be connected, we can choose x^0 in a bounded component of $\mathbf{C}\setminus K$ and note that (1) violates the maximum principle.

Assume that $\mathbf{C}\setminus K$ is connected. Fix $x^0 \in \mathbf{C}\setminus K$. Then $K \cup \{x^0\}$ is a set with connected complement. Choose points $x_n \to x^0$ and $x_n \neq x^0$. Then

$$f_n(z) = \frac{1}{z - x_n}$$

is holomorphic in a neighborhood of $K \cup \{x^0\}$. Hence by Theorem 7.1 we can find a polynomial P_n with

$$\left| P_n(z) - \frac{1}{z - x_n} \right| < \frac{1}{n}, \qquad \text{all } z \in K \cup \{x_0\}.$$

For large n, then, P_n satisfies (1). Q.E.D.

Definition 7.1. Let X be a compact subset of $\mathbf{C}^n$. We define the *polynomially convex hull of* X, denoted $h(X)$, by

$$h(X) = \{z \in \mathbf{C}^n \,|\, |Q(z)| \leq \max_X |Q|$$
for every polynomial $Q\}$.

Evidently $h(X)$ is a compact set containing X.

Definition 7.2. X is said to be *polynomially convex* if $h(X) = X$.

Note that X is polynomially convex if and only if for every x^0 in $\mathbf{C}^n\setminus X$ we can find a polynomial P with

$$|P(x^0)| > \max_X |P|. \tag{2}$$

For $X \subset \mathbf{C}$, Lemma 7.2 now gives that $\mathbf{C} \setminus X$ is connected if and only if X is polynomially convex. Theorem 7.1 can now be stated: For $X \subset \mathbf{C}$, the approximation problem on X is solvable provided that X is polynomially convex. Formulated in this way, the theorem admits generalization to $\mathbf{C}^n$ for $n > 1$.

THEOREM 7.3 (OKA–WEIL)

Let X be a compact, polynomially convex set in $\mathbf{C}^n$. Then for every function f holomorphic in some neighborhood of X, we can find a sequence $\{P_j\}$ of polynomials in $z_1, \ldots, z_n$ with

$$P_j \to f \text{ uniformly on } X.$$

Note. In order to apply this result in particular cases we of course have to verify that a given set X is polynomially convex. This is usually quite difficult. However, we shall see that in the theory of Banach algebras polynomially convex sets arise in a natural way.

André Weil, who first proved the essential portion of Theorem 7.3 [L'Intégrale de Cauchy et les fonctions de plusieurs variables, *Math. Ann.* **111** (1935), 178–182], made use of a generalization of the Cauchy integral formula to several complex variables. We shall follow another route, due to Oka, based on the Oka extension theorem given below.

Definition 7.3. A subset Π of $\mathbf{C}^n$ is a *p-polyhedron* if there exist polynomials $P_1, \ldots, P_s$ such that

$$\Pi = \{z \in \mathbf{C}^n \mid |z_j| \leq 1, \text{ all } j, \text{ and } |P_k(z)| \leq 1,\ k = 1, 2, \ldots, s\}.$$

LEMMA 7.4

Let X be a compact polynomially convex subset of Δ^n. Let $\mathcal{O}$ be an open set containing X. Then there exists a p-polyhedron Π with $X \subset \Pi \subset \mathcal{O}$.

Proof. For each $x \in \Delta^n \setminus \mathcal{O}$ there exists a polynomial P_x with $|P_x(x)| > 1$ and $|P_x| \leq 1$ on X.

Then $|P_x| > 1$ in some neighborhood $\mathcal{N}_x$ of x. By compactness of $\Delta^n \setminus \mathcal{O}$, a finite collection $\mathcal{N}_{x_1}, \ldots, \mathcal{N}_{x_r}$ covers $\Delta^n \setminus \mathcal{O}$. Put

$$\Pi = \{z \in \Delta^n \mid |P_{x_1}(z)| \leq 1, \ldots, |P_{x_r}(z)| \leq 1\}.$$

If $z \in X$, then $z \in \Pi$, so $X \subset \Pi$.

Suppose that $z \notin \mathcal{O}$. If $z \notin \Delta^n$, then $z \notin \Pi$. If $z \in \Delta^n$, then $z \in \Delta^n \setminus \mathcal{O}$. Hence $z \in \mathcal{N}_{x_j}$ for some j. Hence $|P_{x_j}(z)| > 1$. Thus $z \notin \Pi$. Hence $\Pi \subset \mathcal{O}$. Q.E.D.

Let now Π be a p-polyhedron in $\mathbf{C}^n$,

$$\Pi = \{z \in \Delta^n \mid |P_j(z)| \leq 1,\ j = 1, \ldots, r\}.$$

We can embed Π in $\mathbf{C}^{n+r}$ by the map

$$\Phi : z \to (z, P_1(z), \ldots, P_r(z)).$$

Φ maps Π homeomorphically onto the subset of Δ^{n+r} defined by the equations

$$z_{n+1} - P_1(z) = 0, \dots, z_{n+r} - P_r(z) = 0.$$

THEOREM 7.5 (OKA EXTENSION THEOREM)

Given f holomorphic in some neighborhood of Π; then there exists F holomorphic in a neighborhood of Δ^{n+r} such that

$$F(z, P_1(z), \dots, P_r(z)) = f(z), \qquad \text{all } z \in \Pi.$$

The Oka–Weil theorem is an easy corollary of this result.

Proof of Theorem 7.3. Without loss of generality we may assume that $X \subset \Delta^n$. (Why?) f is holomorphic in a neighborhood $\mathcal{O}$ of X. By Lemma 7.4 there exists a p-polyhedron Π with $X \subset \Pi \subset \mathcal{O}$. Then f is holomorphic in a neighborhood of Π. By Theorem 7.5 we can find F satisfying

(3) $$F(z, P_1(z), \dots, P_r(z)) = f(z), \qquad z \in \Pi,$$

F holomorphic in a neighborhood of Δ^{n+r}. Expand F in a Taylor series around 0,

$$F(z, z_{n+1}, \dots, z_{n+r}) = \sum_{\nu} a_\nu z_1^{\nu_1} \cdots z_n^{\nu_n} z_{n+1}^{\nu_{n+1}} \cdots z_{n+r}^{\nu_{n+r}}.$$

The series converges uniformly in Δ^{n+r}. Thus a sequence $\{S_j\}$ of partial sums of this series converges uniformly to F on Δ^{n+r}, and hence in particular on $\Phi(\Pi)$. Thus

$$S_j(z, P_1(z), \dots, P_r(z))$$

converges uniformly to $F(z, P_1(z), \dots, P_r(z))$ for $z \in \Pi$, or, in other words, converges to $f(z)$, by (3). Since $S_j(z, P_1(z), \dots, P_r(z))$ is a polynomial in z for each j, we are done.

We must now attack the Oka Extension theorem. We begin with a generalization of Theorem 6.1.

THEOREM 7.6

Let Π be a p-polyhedron in $\mathbf{C}^n$ and Ω a neighborhood of Π. Given that $\phi \in \Lambda^{p,q}(\Omega)$, $q > 0$, with $\bar{\partial}\phi = 0$, then there exists a neighborhood Ω_1 of Π and $\psi \in \Lambda^{p,q-1}(\Omega_1)$ with $\bar{\partial}\psi = \phi$.

First we need some definitions and exercises.

Let Ω be an open set in $\mathbf{C}^n$ and W an open set in $\mathbf{C}^k$. Let $u = (u_1, \dots, u_n)$ be a map of W into Ω. Assume that each $u_j \in C^\infty(W)$.

Exercise 7.1. Let $a \in C^\infty(\Omega)$, so $a(u) \in C^\infty(W)$. Then

$$d\{a(u)\} = \sum_{j=1}^{n} \frac{\partial a}{\partial z_j}(u)\, du_j + \frac{\partial a}{\partial \bar{z}_j}(u)\, d\bar{u}_j.$$

Both sides are forms in $\Lambda^1(W)$.

Let Ω, W, and u be as above. Assume that each $u_j \in H(W)$. For each $I = (i_1, \dots, i_r)$, $J = (j_1, \dots, j_s)$ put

$$du_I = du_{i_1} \wedge du_{i_2} \wedge \cdots \wedge du_{i_r}$$

and define $d\bar{u}_J$ similarly. Thus $d\bar{u}_I \wedge du_J \in \Lambda^{r,s}(W)$.

Fix $\omega \in \Lambda^{r,s}(\Omega)$,

$$\omega = \sum_{I,J} a_{IJ}\, dz_I \wedge d\bar{z}_J.$$

Definition 7.4

$$\omega(u) = \sum_{I,J} a_{IJ}(u)\, du_I \wedge d\bar{u}_J \in \Lambda^{r,s}(W).$$

Exercise 7.2. $d(\omega(u)) = (d\omega)(u)$ and $\bar{\partial}(\omega(u)) = (\bar{\partial}\omega)(u)$. We still assume, in this exercise, that each u_j is holomorphic.

Proof of Theorem 7.6. We denote

$$P^k(q_1, \ldots, q_r) = \{z \in \Delta^k \,|\, |q_j(z)| \leq 1, j = 1, \ldots, r\},$$

the q_j being polynomials in $z_1, \ldots, z_k$. Every p-polyhedron is of this form.

We shall prove our theorem by induction on r. The case $r = 0$ corresponds to the p-polyhedron Δ^k and the assertion holds, for all k, by Theorem 6.1.

Fix r now and suppose that the assertion holds for this r and all k and all (p, q), $q > 0$. Fix n and polynomials $p_1, \ldots, p_{r+1}$ in $\mathbf{C}^n$ and consider $\phi \in \Delta^{p,q}(\Omega)$, Ω some neighborhood of $P^n(p_1, \ldots, p_{r+1})$. We first sketch the argument.

Step 1. Embed $P^n(p_1, \ldots, p_{r+1})$ in $P^{n+1}(p_1, \ldots, p_r)$ by the map $u: z \to (z, p_{r+1}(z))$. Note that $p_1, \ldots, p_r$ are polynomials in $z_1, \ldots, z_{n+1}$ which do not involve z_{n+1}. Let Σ denote the image of $P^n(p_1, \ldots, p_{r+1})$ under u. π denotes the projection $(z, z_{n+1}) \to z$ from $\mathbf{C}^{n+1} \to \mathbf{C}^n$. Note $\pi \circ u =$ identity.

Step 2. Find a $\bar{\partial}$-closed form Φ_1 defined in a neighborhood of $P^{n+1}(p_1, \ldots, p_r)$ with $\Phi_1 = \phi(\pi)$ on Σ.

Step 3. By induction hypothesis, $\exists \Psi$ in a neighborhood of $P^{n+1}(p_1, \ldots, p_r)$ with $\bar{\partial}\Psi = \Phi_1$. Put $\psi = \Psi(u)$. Then

$$\bar{\partial}\psi = (\bar{\partial}\Psi)(u) = \Phi_1(u) = \phi.$$

As to the details, choose a neighborhood Ω_1 of $P^n(p_1, \ldots, p_{r+1})$ with $\bar{\Omega}_1 \subset \Omega$. Choose $\lambda \in C^\infty(\mathbf{C}^n)$, $\lambda = 1$ on $\bar{\Omega}_1$, $\lambda = 0$ outside Ω. Put $\Phi = (\lambda \cdot \phi)(\pi)$, defined $= 0$ outside $\pi^{-1}(\Omega)$.

Let χ be a form of type (p, q) defined in a neighborhood of $P^{n+1}(p_1, \ldots, p_r)$. Put

$$\Phi_1 = \Phi - (z_{n+1} - p_{r+1}(z)) \cdot \chi. \tag{4}$$

Then $\Phi_1 = \Phi = \phi(\pi)$ on Σ.

We want to choose χ such that Φ_1 is $\bar{\partial}$-closed. This means that

$$\bar{\partial}\Phi = (z_{n+1} - p_{r+1}(z))\bar{\partial}\chi$$

or

$$\bar{\partial}\chi = \frac{\bar{\partial}\Phi}{(z_{n+1} - p_{r+1}(z))}. \tag{5}$$

Observe that $\bar{\partial}\Phi = \bar{\partial}\phi(\pi) = 0$ in a neighborhood of Σ, whence the right-hand side in (5) can be taken to be 0 in a neighborhood of Σ and is then in C^∞ in a neighborhood of $P^{n+1}(p_1, \ldots, p_r)$. Also

$$\bar{\partial}\left\{\frac{\bar{\partial}\Phi}{z_{n+1} - p_{r+1}(z))}\right\} = 0.$$

By induction hypothesis, now, $\exists\chi$ satisfying (5). The corresponding Φ_1 in (4) is then $\bar{\partial}$-closed in some neighborhood of $P^{n+1}(p_1, \ldots, p_r)$. By induction hypothesis again, $\exists a$ $(p, q-1)$ form Ψ in a neighborhood of $P^{n+1}(p_1, \ldots, p_r)$ with $\bar{\partial}\Psi = \Phi_1$. As in step 3, then, making use of Exercise 7.2, we obtain a $(p, q-1)$ form ψ in a neighborhood of $P^n(p_1, \ldots, p_{r+1})$ with $\bar{\partial}\psi = \phi$. Q.E.D.

We keep the notations introduced in the last proof.

LEMMA 7.7

Fix k and polynomials $q_1, \ldots, q_r$ in $z = (z_1, \ldots, z_k)$. Let f be holomorphic in a neighborhood W of $\Pi = P^k(q_1, \ldots, q_r)$. Then $\exists F$ holomorphic in a neighborhood of $\Pi' = P^{k+1}(q_2, \ldots, q_r)$ such that

$$F(z, q_1(z)) = f(z), \qquad \text{all } z \in \Pi.$$

[Note that if $z \in \Pi$, then $(z, q_1(z)) \in \Pi'$.]

Proof. Let Σ be the subset of Π' defined by $z_{k+1} - q_1(z) = 0$. Choose $\phi \in C_0^\infty(\pi^{-1}(W))$ with $\phi = 1$ in a neighborhood of Σ.

We seek a function G defined in a neighborhood of Π' so that with

$$F(z, z_{k+1}) = \phi(z, z_{k+1})f(z) - (z_{k+1} - q_1(z))G(z, z_{k+1}),$$

F is holomorphic in a neighborhood of Π'. We define $\phi \cdot f = 0$ outside $\pi^{-1}(W)$. We need $\bar{\partial}F = 0$ and so

$$f\bar{\partial}\phi = (z_{k+1} - q_1(z))\bar{\partial}G$$

or

$$\bar{\partial}G = \frac{f\bar{\partial}\phi}{(z_{k+1} - q_1(z))} = \omega. \tag{6}$$

Note that the numerator vanishes in a neighborhood of Σ, so ω is a smooth form in some neighborhood of Π'. Also $\bar{\partial}\omega = 0$. By Theorem 7.6, we can thus find G satisfying (6) in some neighborhood of Π'. The corresponding F now has the required properties. Q.E.D.

Proof of Theorem 7.5. $p_1, \ldots, p_r$ are given polynomials in $z_1, \ldots, z_n$ and $\Pi = P^n(p_1, \ldots, p_r)$. f is holomorphic in a neighborhood of Π. For $j = 1, 2, \ldots, r$ we consider the assertion

$A(j)$: $\exists F_j$ holomorphic in a neighborhood of $P^{n+j}(p_{j+1}, \ldots, p_r)$

such that $F_j(z, p_1(z), \ldots, p_j(z)) = f(z)$, all $z \in \Pi$.

$A(1)$ holds by Lemma 7.7. Assume that $A(j)$ holds for some j. Thus F_j is holomorphic in a neighborhood of $P^{n+j}(p_{j+1}, \ldots, p_r)$. By Lemma 7.7, $\exists F_{j+1}$ is holomorphic in a neighborhood of $P^{n+j+1}(p_{j+2}, \ldots, p_r)$ with $F_{j+1}(\zeta, p_{j+1}(z)) = F_j(\zeta)$, $\zeta \in P^{n+j}(p_{j+1}, \ldots, p_r)$, and $\zeta = (z, z_{n+1}, \ldots, z_{n+j})$.

By choice of F_j.

$$F_j(z, p_1(z), \ldots, p_j(z)) = f(z), \qquad \text{all } z \text{ in } \Pi.$$

Hence

$$F_{j+1}(z, p_1(z), \ldots, p_j(z), p_{j+1}(z)) = f(z), \qquad \text{all } z \text{ in } \Pi.$$

Thus $A(j+1)$ holds. Hence $A(1), A(2), \ldots, A(r)$ all hold. But $A(r)$ provides F holomorphic in a neighborhood of Δ^{n+r} with

$$F(z, p_1(z), \ldots, p_r(z)) = f(z), \qquad \text{all } z \text{ in } \Pi. \qquad \text{Q.E.D.}$$

Exercise 7.3. Let $\mathfrak{A}$ be a uniform algebra on a compact space X with generators $g_1, \ldots, g_n$ (i.e., $\mathfrak{A}$ is the smallest closed subalgebra of itself containing the g_j). Show that the map

$$x \to (\hat{g}_1(x), \ldots, \hat{g}_n(x))$$

maps $\mathcal{M}(\mathfrak{A})$ onto a compact, polynomially convex set K in $\mathbf{C}^n$, and that this map carries $\mathfrak{A}$ isomorphically and isometrically onto $P(K)$.

Exercise 7.4. Let X be a compact set in $\mathbf{C}^n$. Show that $\mathcal{M}(P(X))$ can be identified with $h(X)$. In particular, if X is polynomially convex, $\mathcal{M}(P(X)) = X$.

NOTES

Theorem 7.5 and the proof of Theorem 7.3 based on it is due to K. Oka, Domaines convexes par rapport aux fonctions rationelles, *J. Sci. Hiroshima Univ.* **6** (1936), 245–255. The proof of Theorem 7.5 given here is found in Gunning and Rossi [6, Chap. 1].

8

OPERATIONAL CALCULUS IN SEVERAL VARIABLES

We wish to extend the operational calculus established in Section 3 to functions of several variables. Let $\mathfrak{A}$ be a Banach algebra and $x_1, \ldots, x_n \in \mathfrak{A}$. If P is a polynomial in n variables

$$P(z_1, \ldots, z_n) = \sum_{\nu} A_\nu z_1^{\nu_1} \cdots z_n^{\nu_n},$$

it is natural to define

$$P(x_1, \ldots, x_n) = \sum_{\nu} A_\nu x_1^{\nu_1} \cdots x_n^{\nu_n} \in \mathfrak{A}.$$

We then observe that if $y = P(x_1, x_2, \ldots, x_n)$, then

(1) $$\hat{y} = P(\hat{x}_1, \ldots, \hat{x}_n) \text{ on } \mathcal{M}.$$

Let F be a complex-valued function defined on an open set $\Omega \subset \mathbf{C}^n$. In order to define $F(\hat{x}_1, \ldots, \hat{x}_n)$ on $\mathcal{M}$ we must assume that Ω contains

$$\{(\hat{x}_1(M), \ldots, \hat{x}_n(M)) | M \in \mathcal{M}\}.$$

Definition 8.1. *$\sigma(x_1, \ldots, x_n)$, the joint spectrum of $x_1, \ldots, x_n$, is $\{(\hat{x}_1(M), \ldots, \hat{x}_n(M)) | M \in \mathcal{M}\}$.*

When $n = 1$, we recover the old spectrum $\sigma(x)$. You easily verify

LEMMA 8.1

$(\lambda_1, \ldots, \lambda_n)$ in $\mathbf{C}^n$ lies in $\sigma(x_1, \ldots, x_n)$ if and only if the equation

$$\sum_{j=1}^{n} y_j(x_j - \lambda_j) = 1$$

has no solution $y_1, \ldots, y_n \in \mathfrak{A}$.

We shall prove

THEOREM 8.2

Fix $x_1, \ldots, x_n \in \mathfrak{A}$. *Let* Ω *be an open set in* $\mathbf{C}^n$ *with* $\sigma(x_1, \ldots, x_n) \subset \Omega$. *For each* $F \in H(\Omega)$ *there exists* $y \in \mathfrak{A}$ *with*

$$\hat{y}(M) = F(\hat{x}_1(M), \ldots, \hat{x}_n(M)), \qquad \text{all } M \in \mathcal{M}. \tag{2}$$

Remark. This result is, of course, not a full generalization of Theorem 3.3. We shall see that it is adequate for important applications, however. When $\mathfrak{A}$ is semisimple, we can say more. In that case y is determined uniquely by (2) and we can define

$$F(x_1, \ldots, x_n) = y.$$

Now $H(\Omega)$ is an F-space in the sense of [4, Chap. II]. Hence by the closed graph theorem (*loc. cit.*), the map

$$F \to F(x_1, \ldots, x_n)$$

is continuous from $H(\Omega) \to \mathfrak{A}$. Thus

COROLLARY

If $\mathfrak{A}$ *is semisimple,* $F_j \to F$ *in* $H(\Omega)$ *implies that* $F_j(x_1, \ldots, x_n) \to F(x_1, \ldots, x_n)$ *in* $\mathfrak{A}$.

We shall first prove our theorem under the assumption that

(3) $x_1, \ldots, x_n$ generate $\mathfrak{A}$; i.e., the smallest closed subalgebra of $\mathfrak{A}$ containing $x_1, \ldots, x_n$ coincides with $\mathfrak{A}$.

LEMMA 8.3

Assume (3). *Then* $\sigma(x_1, \ldots, x_n)$ *is a polynomially convex subset of* $\mathbf{C}^n$.

Proof. Fix $z^0 = (z_1^0, \ldots, z_n^0)$ with

$$|Q(z^0)| \leq \max_\sigma |Q|, \qquad \text{all polynomials } Q,$$

where $\sigma = \sigma(x_1, \ldots, x_n)$.

$$\max_\sigma |Q| = \max_{\mathcal{M}} |Q(\hat{x}_1, \ldots, \hat{x}_n)| = \max_{\mathcal{M}} |\widehat{Q(x_1, \ldots, x_n)}|$$
$$\leq \|Q(x_1, \ldots, x_n)\|.$$

Hence the map $\chi : Q(x_1, \ldots, x_n) \to Q(z^0)$ is a bounded homomorphism from a dense subalgebra of $\mathfrak{A} \to \mathbf{C}$. (Check that χ is unambiguously defined.) Hence χ extends to a homomorphism of $\mathfrak{A} \to \mathbf{C}$, so $\exists M_0 \in \mathcal{M}$ with $\chi(f) = \hat{f}(M_0)$, all $f \in \mathfrak{A}$. In particular,

$$\chi(x_j) = \hat{x}_j(M_0) \quad \text{or} \quad z_j^0 = \hat{x}_j(M_0), \qquad j = 1, \ldots, n.$$

Thus $z^0 \in \sigma$. Hence σ is polynomially convex. Q.E.D.

Exercise 8.1. Let F be holomorphic in a neighborhood of Δ^N with

$$F(\zeta) = \sum_{\nu} C_\nu \zeta_1^{\nu_1} \cdots \zeta_N^{\nu_N}.$$

Given that $y_1, \ldots, y_N \in \mathfrak{A}$, $\max_{\mathcal{M}} |\hat{y}_j| \leq 1$, all j. Then

$$\sum_{\nu} C_\nu y_1^{\nu_1} \cdots y_N^{\nu_N}$$

converges in $\mathfrak{A}$.

Proof of Theorem 8.2, assuming (3). Without loss of generality, $\|x_j\| \leq 1$ for all j. By Lemma 8.3, $\sigma = \sigma(x_1, \ldots, x_n)$ is polynomially convex, and $\sigma \subset \Delta^n$. By Lemma 7.4, $\exists$ a p-polyhedron Π with $\sigma \subset \Pi \subset \Omega$, $\Pi = P^n(p_1, \ldots, p_r)$. Fix $\phi \in H(\Omega)$. By the Oka extension theorem, $\exists\Phi$ holomorphic in a neighborhood of Δ^{n+r} with

$$\Phi(z_1, \ldots, z_n, p_1(z), \ldots, p_r(z)) = \phi(z), \qquad z \in \Pi.$$

Put $y_1 = x_1, \ldots, y_n = x_n, y_{n+1} = p_1(x_1, \ldots, x_n), \ldots, y_{n+r} = p_r(x_1, \ldots, x_n)$. We verify that $\max_{\mathcal{M}} |\hat{y}_j| \leq 1, j = 1, 2, \ldots, n + r$. By Exercise 8.1,

$$\sum_{\nu} C_\nu x_1^{\nu_1} \cdots c_n^{\nu_n} (p_1(x))^{\nu_{n+1}} \cdots (p_r(x))^{\nu_{n+r}}$$

converges in $\mathfrak{A}$ to an element y, where $\Sigma_\nu C_\nu \zeta^\nu$ is the Taylor expansion of Φ at 0 and $p_j(x)$ denotes $p_j(x_1, \ldots, x_n)$. Then

$$\begin{aligned} \hat{y}(M) &= \Phi(\hat{x}_1(M), \ldots, \hat{x}_n(M), p_1(\hat{x}(M)), \ldots, p_r(\hat{x}(M))) \\ &= \phi(\hat{x}_1(M), \ldots, \hat{x}_n(M)), \qquad \text{all } M \in \mathcal{M}, \end{aligned}$$

since $(\hat{x}_1(M), \ldots, \hat{x}_n(M)) \in \sigma \subset \Pi$. We are done.

If we now drop (3), σ is no longer polynomially convex. Richard Arens and Alberto Calderon fortunately found a way to reduce the general case to the finitely generated one.

Let $x_1, \ldots, x_n \in \mathfrak{A}$, let W be an open set in $\mathbf{C}^n$ containing $\sigma(x_1, \ldots, x_n)$, and fix $F \in H(W)$. For every closed subalgebra $\mathfrak{A}'$ of $\mathfrak{A}$ containing elements $\xi_1, \ldots, \xi_k$ of $\mathfrak{A}$, let $\sigma_{\mathfrak{A}'}(\xi_1, \ldots, \xi_k)$ denote the joint spectrum of $\xi_1, \ldots, \xi_k$ relative to $\mathfrak{A}'$.

Assertion. $\exists C_1, \ldots, C_m \in \mathfrak{A}$ such that if B is the closed subalgebra of $\mathfrak{A}$ generated by $x_1, \ldots, x_n, C_1, \ldots, C_m$, then

(4) $$\sigma_B(x_1, \ldots, x_n) \subset W.$$

Grant this for now. Let π be the projection $(z_1, \ldots, z_n, z_{n+1}, \ldots, z_{n+m}) \to (z_1, \ldots, z_n)$ of $\mathbf{C}^{n+m} \to \mathbf{C}^n$. Because of (4), $\sigma_B(x_1, \ldots, x_n, C_1, \ldots, C_m) \subset \pi^{-1}(W)$. Define a function ϕ on $\pi^{-1}(W)$ by

$$\phi(z_1, \ldots, z_n, z_{n+1}, \ldots, z_{n+m}) = F(z_1, \ldots, z_n).$$

Thus ϕ is holomorphic in a neighborhood of $\sigma_B(x_1, \ldots, x_n, C_1, \ldots, C_m)$, and so, by Theorem 8.2 under hypothesis (3) applied to B and the set of generators $x_1, \ldots, C_m$, $\exists y \in B$ with

$$\begin{aligned}\hat{y} &= \phi(\hat{x}_1, \ldots, \hat{x}_n, \hat{C}_1, \ldots, \hat{C}_m)\\ &= F(\hat{x}_1, \ldots, \hat{x}_n) \text{ on } \mathcal{M}(B).\end{aligned}$$

If $M \in \mathcal{M}$, then $M \cap B \in \mathcal{M}(B)$ and hence $\hat{y}(M) = F(\hat{x}_1(M), \ldots, \hat{x}_n(M))$. We are done, except for the proof of the assertion.

Let $\mathfrak{A}_0$ denote the closed subalgebra generated by $x_1, \ldots, x_n$ and put $\sigma_0 = \sigma_{\mathfrak{A}_0}(x_1, \ldots, x_n)$. If $\sigma_0 \subset W$, take $B = \mathfrak{A}_0$. If not, consider $\zeta \in \sigma_0 \setminus W$.

Since $\zeta \notin \sigma(x_1, \ldots, x_n)$, $\exists y_1, \ldots, y_n \in \mathfrak{A}$ such that $\Sigma_{j=1}^n y_j(x_j - \zeta_j) = 1$. Denote by $\mathfrak{A}(\zeta)$ the closed subalgebra generated by $x_1, \ldots, x_n, y_1, \ldots, y_n$. Then $\exists$ neighborhood $\mathcal{N}_\zeta$ of ζ in $\mathbf{C}$ such that if $\alpha \in \mathcal{N}_\zeta$, then $\Sigma_j y_j(x_j - \alpha_j)$ is invertible in $\mathfrak{A}(\zeta)$. It follows that if $\alpha \in \mathcal{N}_\zeta$, then $\alpha \notin \sigma_{\mathfrak{A}(\zeta)}(x_1, \ldots, x_n)$.

By compactness of $\sigma_0 \setminus W$, we obtain in this way a finite covering of $\sigma_0 \setminus W$ by neighborhoods $\mathcal{N}_\zeta$. We throw together all the corresponding y_j and call them $C_1, \ldots, C_m$, and we let B be the closed subalgebra generated by $x_1, \ldots, x_n, C_1, \ldots, C_m$. Note that $\sigma_B(x_1, \ldots, x_n) \subset \sigma_0$. (Why?) If $\alpha \in \sigma_0 \setminus W$, then α lies in one of our finitely many $\mathcal{N}_\zeta$, and so $\exists u_1, \ldots, u_n \in B$ such that $\Sigma_j u_j(x_j - \alpha_j)$ is invertible in B. Hence $\alpha \notin \sigma_B(x_1, \ldots, x_n)$. Thus $\sigma_B(x_1, \ldots, x_n) \subset W$, proving the assertion. Thus Theorem 8.2 holds in general.

As a first application we consider this problem. Let $\mathfrak{A}$ be a Banach algebra and $x \in \mathfrak{A}$. When does x have a square root in $\mathfrak{A}$, i.e., when can we find $y \in \mathfrak{A}$ with $y^2 = x$?

An obvious necessary condition is the purely topological one:

$$\exists y \in C \qquad \text{with } y^2 = \hat{x} \text{ on } \mathcal{M}. \tag{5}$$

Condition (5) alone is not sufficient, as is seen by taking, with $D = \{z \,|\, |z| \leq 1\}$,

$$\mathfrak{A} = \{f \in A(D) \,|\, f'(0) = 0\}.$$

Then $z^2 \in \mathfrak{A}$, $z \notin \mathfrak{A}$, but (5) holds. However, one can prove

THEOREM 8.4

Let $\mathfrak{A}$ be a Banach algebra, $a \in \mathfrak{A}$. and assume that $\exists h \in C(\mathcal{M})$ with $h^2 = \hat{a}$. Assume also that $\hat{a}$ never vanishes on $\mathcal{M}$. Then a has a square root in $\mathfrak{A}$.

We approach the proof as follows: First find $a_2, \ldots, a_n \in \mathfrak{A}$ such that $\exists F$ holomorphic in a neighborhood of $\sigma(a, a_2, \ldots, a_n)$ in $\mathbf{C}^n$ with $F^2 = z_1$. By Theorem 8.2, $\exists y \in \mathfrak{A}$, with $\hat{y} = F(\hat{a}, \hat{a}_2, \ldots, \hat{a}_n)$ on $\mathcal{M}$. Then $\hat{y}^2 = \hat{a}$ on $\mathcal{M}$. If $\mathfrak{A}$ is semisimple, we are done. In the general case, put $\rho = a - y^2$. Then $\rho \in \operatorname{rad} \mathfrak{A}$. Since $\hat{y}^2 = \hat{a}$, y^2 is invertible and $\rho/y^2 \in \operatorname{rad} \mathfrak{A}$. Then $(y\sqrt{1 + \rho/y^2})^2 = y^2(1 + \rho/y^2) = a$, so $y\sqrt{1 + \rho/y^2}$ solves our problem provided that $\sqrt{1 + \rho/y^2} \in \mathfrak{A}$. It does so by

Exercise 8.2. Let $\mathfrak{A}$ be a Banach algebra and $x \in \operatorname{rad} \mathfrak{A}$. Then $\exists \zeta \in \mathfrak{A}$ with $\zeta^2 = 1 + x$ and $\hat{\zeta} \equiv 1$ on $\mathcal{M}$.

We return to the details.

LEMMA 8.5

Given a as in Theorem 8.4, $\exists a_2, \ldots, a_n \in \mathfrak{A}$ *such that if* $K = \sigma(a, a_2, \ldots, a_n) \subset \mathbf{C}^n$, *then we can find* $H \in C(K)$ *with* $H^2 = z_1$ *on* K.

Proof. In the topological product $\mathcal{M} \times \mathcal{M}$ put $S = \{(M, M') | h(M) + h(M') = 0\}$, where h is as in Theorem 8.4. S is compact and disjoint from the diagonal. (Why?) Let $x = (M_1, M'_1) \in S$. Since $M'_1 \neq M_1$, $\exists b_x \in \mathfrak{A}$ with $\widehat{b_x}(M_1) - \widehat{b_x}(M'_1) \neq 0$. By continuity $\widehat{b_x}(M) - \widehat{b_x}(M') \neq 0$ for all (M, M') in some neighborhood $\mathcal{N}_x$ of x in S. By compactness, $\mathcal{N}_{x_2}, \ldots, \mathcal{N}_{x_n}$ cover S for a suitable choice of $x_2, \ldots, x_n$. Put $a_j = b_{x_j}, j = 2, \ldots, n$. Put $K = \sigma(a, a_2, \ldots, a_n)$ and fix $z = (\hat{a}(M), \hat{a}_2(M), \ldots, \hat{a}_n(M)) \in K$.

We define a function H on K by $H(z) = h(M)$. To see that H is well defined, suppose that for $(M, M') \in \mathcal{M} \times \mathcal{M}$,

$$\hat{a}(M) = \hat{a}(M'), \quad \hat{a}_j(M) = \hat{a}_j(M'), \qquad j = 2, \ldots, n. \tag{6}$$

$(M, M') \notin S$, for this would imply that $(M, M') \in \mathcal{N}_{x_j}$ for some j, denying (6). Hence $h(M) \neq -h(M')$. By (6), $h^2(M) = h^2(M')$. Hence $h(M) = h(M')$, as desired. It is easily verified that H is continuous on K, and that $H^2 = z_1$. Q.E.D.

Proof of Theorem 8.4. It only remains to construct F holomorphic in a neighborhood of K with $F^2 = z_1$.

For each $x \in K$ and $r > 0$, let $B(x, r)$ be the open ball in $\mathbf{C}^n$ centered at x and of radius r. If $x = (\alpha_1, \ldots, \alpha_n) \in K$, $\alpha_1 \neq 0$. Hence $\exists r > 0$ and F_x holomorphic in $B(x, r)$, with $F_x^2 = z_1$ in $B(x, r)$. By compactness of K, a fixed r will work for all x in K. This is not enough, however, to yield an F holomorphic in a neighborhood of K with $F^2 = z_1$. (Why not?) But we can require, in addition, that $F_x = H$ in $B(x, r) \cap K$. Put $\Omega = \bigcup_{x \in K} B(x, r/2)$. For $\zeta \in \Omega$, define $F(\zeta) = F_x(\zeta)$ if $\zeta \in B(x, r/2)$, $x \in K$. To see that this value is independent of $x \in K$, suppose that $\zeta \in B(x, r/2) \cap B(y, r/2)$, $x, y \in K$.

Then $y \in B(x, r) \cap K$. Hence $F_x(y) = H(y)$. Also, $F_y(y) = H(y)$. Hence F_x and F_y are two holomorphic functions in $B(x, r) \cap B(y, r/2)$ with $F_x^2 = F_y^2 = z_1$ there and $F_x = F_y$ at y. So $F_x(\zeta) = F_y(\zeta)$. (Why?) Thus $F \in H(\Omega)$ and $F^2 = z_1$ in Ω. Q.E.D.

Theorem 8.4 holds when the square-root function is replaced by any one of a large class of multivalued analytic functions. (See the Notes at the end of this section.)

As our second application of Theorem 8.2, we take the existence of idempotent elements.

THEOREM 8.6 (ŠILOV IDEMPOTENT THEOREM)

Let $\mathfrak{A}$ *be a Banach algebra and assume that* $\mathcal{M} = \mathcal{M}_1 \cup \mathcal{M}_2$, *where* $\mathcal{M}_1$ *and* $\mathcal{M}_2$ *are disjoint closed sets. Then* $\exists e \in \mathfrak{A}$ *with* $e^2 = e$ *and* $\hat{e} = 1$ *on* $\mathcal{M}_1$ *and* $\hat{e} = 0$ *on* $\mathcal{M}_2$.

LEMMA 8.7

$\exists a_1, \ldots, a_N \in \mathfrak{A}$ *such that if* $\hat{a}$ *is the map of* $\mathcal{M} \to C^N : M \to (\hat{a}_1(M), \ldots, \hat{a}_N(M))$, *then* $\hat{a}(\mathcal{M}_1) \cap \hat{a}(\mathcal{M}_2) = \varnothing$.

The proof is like that of Lemma 8.5 and is left to the reader.

Proof of Theorem 8.6. By the last Lemma, $\exists a_1, \ldots, a_N \in \mathfrak{A}$, so that $\hat{a}(\mathcal{M}_1)$ and $\hat{a}(\mathcal{M}_2)$ are disjoint compact subsets of $\mathbf{C}^N$. Choose disjoint open sets W_1 and W_2 in

$\mathbf{C}^N$ with $\hat{a}(\mathcal{M}_j) \subset W_j, j = 1, 2$. Put $W = W_1 \cup W_2$ and define F in W by $F = 1$ on W_1 and $F = 0$ on W_2. Then $F \in H(W)$. By Theorem 8.2, $\exists y \in \mathfrak{A}$ with $\hat{y} = F(\hat{a}_1, \ldots, \hat{a}_n)$ on $\mathcal{M}$. Then $\hat{y} = 1$ on $\mathcal{M}_1$, $\hat{y} = 0$ on $\mathcal{M}_2$. We seek $u \in \text{Rad}\, \mathfrak{A}$ so that $(y + u)^2 = y + u$. Then $e = y + u$ will be the desired element.

The condition on $u \Leftrightarrow$

$$u^2 + (2y - 1)u + \rho = 0, \tag{7}$$

where $\rho = y^2 - y \in \text{rad}\, \mathfrak{A}$.

The formula for solving a quadratic equation suggests that we set

$$u = -\frac{2y - 1}{2} + \frac{2y - 1}{2}\zeta,$$

where ζ is the element of $\mathfrak{A}$, provided by Exercise 8.2, satisfying

$$\zeta^2 = 1 - \frac{4\rho}{(2y - 1)^2} \qquad \text{and} \qquad \hat{\zeta} \equiv 1.$$

We can then check that u has the required properties, and the proof is complete.

COROLLARY 1

If $\mathcal{M}$ is disconnected, $\mathfrak{A}$ contains a nontrivial idempotent.

COROLLARY 2

Let $\mathfrak{A}$ be a uniform algebra on a compact space X. Assume that $\mathcal{M}$ is totally disconnected. Then $\mathfrak{A} = C(X)$.

Note. The hypothesis is on $\mathcal{M}$, not on X, but it follows that if $\mathcal{M}$ is totally disconnected, then $\mathcal{M} = X$.

Proof of Corollary 2. If $x_1, x_2 \in X, x_1 \neq x_2$, choose an open and closed set $\mathcal{M}_1$ in $\mathcal{M}$ with $x_1 \in \mathcal{M}_1$, $x_2 \notin \mathcal{M}_1$. Put $\mathcal{M}_2 = \mathcal{M} \setminus \mathcal{M}_1$. By Theorem 8.6, $\exists e \in \mathfrak{A}$, $\hat{e} = 1$ on $\mathcal{M}_1$ and $\hat{e} = 0$ on $\mathcal{M}_2$. Thus e is a real-valued function in $\mathfrak{A}$, which separates x_1 and x_2. By the Stone–Weierstrass theorem, we conclude that $\mathfrak{A} = C(X)$.

COROLLARY 3

Let X be a compact subset of $\mathbf{C}^n$. *Assume that X is polynomially convex and totally disconnected. Then $P(X) = C(X)$.*

Proof. The result follows from Corollary 2, together with the fact that $\mathcal{M}(P(X)) = X$.

NOTES

Theorem 8.2 was proved for finitely generated algebras by G. E. Šilov, On the decomposition of a commutative normed ring into a direct sum of ideals, *A.M.S. Transl.* **1** (1955). The proof given here is due to L. Waelbroeck, Le Calcul symbolique dans les algèbres commutatives, *J. Math. Pure Appl.* **33** (1954), 147–186. Theorem 8.2 for the general case was proved by R. Arens and A. Calderon, Analytic functions of several Banach algebra elements, *Ann. Math.* **62** (1955), 204–216. Theorem 8.4 is a special case of a more general result given by Arens and Calderon, *loc. cit.* Theorem 8.6 and its corollaries are due to Šilov, *loc. cit.*

Our proof of Theorem 8.4 has followed Hörmander's book [8, Chap. 3].

For a stronger version of Theorem 8.2 see Waelbroeck, *loc. cit.*, or N. Bourbaki, Théories spectrales, Hermann, Paris, 1967, Chap. 1, Sec. 4.

9

THE ŠILOV BOUNDARY

Let X be a compact space and $\mathscr{F}$ an algebra of continuous complex-valued functions on X which separates the points of X.

Definition 9.1. A *boundary for* $\mathscr{F}$ is a closed subset E of X such that

$$|f(x)| \le \max_E |f|, \qquad \text{all } f \in \mathscr{F}, x \in X.$$

Thus, for example, if D is the closed unit disk in C and $\mathscr{P}$ the algebra of all polynomials in z, restricted to D, then every closed subset of D containing $\{z \mid |z| = 1\}$ is a boundary for $\mathscr{P}$.

THEOREM 9.1

Let X and $\mathscr{F}$ be as above. Let S denote the intersection of all boundaries for $\mathscr{F}$. Then S is a boundary for $\mathscr{F}$.

Note

(a) It is not clear, a priori, that S is nonempty.

(b) S is evidently closed.

(c) It follows from the theorem that S is the smallest boundary, i.e., that S is a boundary contained in every other boundary.

LEMMA 9.2

Fix $x \in X \setminus S$. $\exists$ a neighborhood U of x with the following property: If β is a boundary, then $\beta \setminus U$ is also a boundary.

Proof. $x \notin S$ and so $\exists$ boundary S_0 with $x \notin S_0$. For each $y \in S_0$, choose $f_y \in \mathscr{F}$ with $f_y(x) = 0$, $f_y(y) = 2$.

$\mathcal{N}_y = \{|f_y| > 1\}$ is a neighborhood of y. Then $\exists y_1, \ldots, y_k$ so that $\mathcal{N}_{y_1} \cup \cdots \cup \mathcal{N}_{y_k} \supset S_0$. Write f_j for f_{y_j}. Put

$$U = \{|f_1| < 1, \ldots, |f_k| < 1\}.$$

Then U is a neighborhood of x and $U \cap S_0 = \varnothing$.

Fix a boundary β and suppose that $\beta \setminus U$ fails to be a boundary. Then $\exists f \in \mathcal{F}$, $\max_X |f| = 1$, with $\max_{\beta \setminus U} |f| < 1$.

Assertion. $\exists n$ so that $\max_X |f^n f_i| < 1$, $i = 1, \ldots, k$.

Grant this for now. Since S_0 is a boundary, we can pick $\bar{x} \in S_0$ with $|f(\bar{x})| = 1$. By the assertion, $|f_i(\bar{x})| < 1$, $i = 1, \ldots, k$.

Hence $\bar{x} \in U$, denying $U \cap S_0 = \varnothing$. Thus $\beta \setminus U$ is a boundary, and we are done.

To prove the assertion, fix M with $\max_X |f_i| < M$, $i = 1, \ldots, k$. Choose n so that $(\max_{\beta \setminus U} |f|)^n \cdot M < 1$. Then $|f^n f_i| < 1$ at each point of $\beta \setminus U$ for every i. On U, $|f^n f_i| < 1$ by choice of U. Hence the assertion.

Proof of Theorem 9.1. Let W be an open set containing S. For each $x \in X \setminus W$ construct a neighborhood U_x by Lemma 9.2. $X \setminus W$ is compact, so we can find finitely many such U_x, say $U_1, \ldots, U_r$, whose union covers $X \setminus W$.

X is a boundary. By choice of U_1, $X \setminus U_1$ is a boundary. Hence $(X/U_1) \setminus U_2$ is a boundary, and at last $X^* = X \setminus (U_1 \cup U_2 \cup \cdots \cup U_r)$ is a boundary. But $X^* \subseteq W$. Hence if $f \in \mathcal{F}$, $\max_X |f| \leq \sup_W |f|$. Since W was an arbitrary neighborhood of S, it follows that S is a boundary. (Why?)

Note. What properties of $\mathcal{F}$ were used in the proof?

Let $\mathfrak{A}$ be a Banach algebra. Then $\hat{\mathfrak{A}}$ is an algebra of continuous functions on $\mathcal{M}$, separating points. By Theorem 9.1 $\exists a$ (unique) boundary S for $\hat{\mathfrak{A}}$ which is contained in every boundary.

Definition 9.2. S is called the *Šilov boundary* of $\mathfrak{A}$ and is denoted $\check{S}(\mathfrak{A})$.

Exercise 9.1. Let Ω be a bounded plane region whose boundary consists of finitely many simple closed curves. Then $\check{S}(A(\Omega))$ = topological boundary $\partial\Omega$ of Ω.

Exercise 9.2. Let Y denote the solid cylinder $= \{(z, t) \in \mathbf{C} \times \mathbf{R} \,|\, |z| \leq 1, 0 \leq t \leq 1\}$. Let $\mathfrak{A}(Y) = \{f \in C(Y) |$ for each t, $f(z, t)$ is analytic in $|z| < 1\}$. Then $\check{S}(\mathfrak{A}(Y)) = \{(z, t) \,|\, |z| = 1, 0 \leq t \leq 1\}$.

Exercise 9.3. Let Y be as in Exercise 9.2 and put $\mathcal{L}(Y) = \{f \in C(Y) | f(z, 1)$ is analytic in $|z| < 1\}$. Then $\check{S}(\mathcal{L}(Y)) = Y$.

Exercise 9.4. Let $\Delta^2 = \{(z, w) \in \mathbf{C}^2 \,|\, |z| \leq 1, |w| \leq 1\}$ and $\mathfrak{A}(\Delta^2) = \{f \in C(\Delta^2) | f \in H(\Omega)$, where Ω = interior of $\Delta^2\}$. Show that $\check{S}(\mathfrak{A}(\Delta^2)) = T = \{(z, w) \,|\, |z| = |w| = 1\}$. Note that here the Šilov boundary is a two-dimensional subset of the three-dimensional topological boundary of Δ^2.

Exercise 9.5. Let $B^n = \{z \in \mathbf{C}^n | \Sigma_{i=1}^n |z_i|^2 \leq 1\}$ and $\mathfrak{A}(B^n) = \{f \in C(B^n) | f \in H(\Omega)$, Ω = interior of $B^n\}$. Show that $\check{S}(\mathfrak{A}(B^n))$ = topological boundary of B^n.

Note that in all these examples, as well as in many others arising naturally, the complement $\mathcal{M} \setminus \check{S}(\mathfrak{A})$ of the Šilov boundary in the maximal ideal space is the union of one or many complex-analytic varieties, and the elements of $\hat{\mathfrak{A}}$ are analytic when restricted to these varieties.

We shall study this phenomenon of "analytic structure" in $\mathcal{M} \setminus \check{S}(\mathfrak{A})$ in several later sections.

We now proceed to consider one respect in which elements of $\hat{\mathfrak{A}}$ *act like* analytic functions on $\mathcal{M}\setminus\check{S}(\mathfrak{A})$.

Let Ω be a bounded domain in $\mathbf{C}$. We have

(1) $$\text{For } F \in \mathfrak{A}(\Omega),\ x \in \Omega,\ |F(x)| \le \max_{\partial\Omega}|F|.$$

The analogous inequality for an arbitrary Banach algebra $\mathfrak{A}$ is true by definition: For $f \in \mathfrak{A}$, $x \in \mathcal{M}$,

$$|\hat{f}(x)| \le \max_{\check{S}(\mathfrak{A})}|\hat{f}|.$$

However, we also have a *local* statement for $\mathfrak{A}(\Omega)$. Fix $x \in \Omega$ and let U be a neighborhood of x in Ω. Then

(2) $$\text{For } F \in \mathfrak{A}(\Omega),\ |F(x)| \le \max_{\partial U}|F|.$$

The analogue of (2) for arbitrary Banach algebras is by no means evident. It is, however, true.

THEOREM 9.3 (LOCAL MAXIMUM MODULUS PRINCIPLE)

Let $\mathfrak{A}$ be a Banach algebra and fix $x \in \mathcal{M}\setminus\check{S}(\mathfrak{A})$. Let U be a neighborhood of x with $U \subset \mathcal{M}\setminus\check{S}(\mathfrak{A})$. Then for all $f \in \mathfrak{A}$,

(3) $$|\hat{f}(x)| \le \max_{\partial U}|\hat{f}|.$$

LEMMA 9.4

Let X be a compact, polynomially convex set in $\mathbf{C}^n$ and U_1 and U_2 be open sets in $\mathbf{C}^n$ with $X \subset U_1 \cup U_2$. If $h \in H(U_1 \cap U_2)$, then $\exists$ a neighborhood W of X and $h_j \in H(W \cap U_j)$, $j = 1, 2$, so that

$$h_1 - h_2 = h \text{ in } W \cap U_1 \cap U_2.$$

Proof. Write $X = X_1 \cup X_2$, where X_j is compact and $X_j \subset U_j$, $j = 1, 2$. Choose $f_1 \in C_0^\infty(U_1)$ with $0 \le f_1 \le 1$ and $f_1 = 1$ on X_1. Similarly, choose $f_2 \in C_0^\infty(U_2)$. Then $f_1 + f_2 \ge 1$ on X, and so $f_1 + f_2 > 0$ in a neighborhood V of X. In V define

$$\eta_1 = \frac{f_1}{f_1 + f_2}, \qquad \eta_2 = \frac{f_2}{f_1 + f_2}.$$

Then $\eta_1, \eta_2 \in C^\infty(V)$, $\eta_1 + \eta_2 = 1$ in V, and $\operatorname{supp} \eta_j \subset U_j$, $j = 1, 2$. With no loss of generality, $U_j = U_j \cap V$. Define functions H_j in $C^\infty(U_j)$, $j = 1, 2$ by

$$H_1 = \eta_2 h \text{ in } U_1 \cap U_2, \qquad H_1 = 0 \text{ in } U_1\setminus U_2.$$

$$H_2 = -\eta_1 h \text{ in } U_1 \cap U_2, \quad H_2 = 0 \text{ in } U_2\setminus U_1.$$

Then

$$H_1 - H_2 = (\eta_1 + \eta_2)h = h \text{ in } U_1 \cap U_2.$$

Hence $\bar{\partial} H_1 = \bar{\partial} H_2$ in $U_1 \cap U_2$. Let f be the (0,1)-form in $U_1 \cup U_2$ defined by $f = \bar{\partial} H_1$ in U_1, $f = \bar{\partial} H_2$ in U_2. Then f is $\bar{\partial}$-closed in $U_1 \cup U_2$. We can choose a p-polyhedron Π with $X \subset \Pi \subset U_1 \cup U_2$. By Theorem 7.6, then, $\exists$ a neighborhood W of Π and $F \in C^\infty(W)$ with $\bar{\partial} F = f$ in W.

Put $h_j = H_j - F$ in $U_j \cap W$, $j = 1, 2$. Then $h_1 - h_2 = h$ in $U_1 \cap U_2 \cap W$, and $\bar{\partial} h_j = f - f = 0$ in $U_j \cap W$; so $h_j \in H(U_j \cap W)$, $j = 1, 2$. Q.E.D.

LEMMA 9.5

Let K be a compact set in $\mathbf{C}^N$ and U_1 and U_2 open sets with

(4) $$U_1 \cup U_2 \supset K,$$

(5) $$U_1 \cap U_2 \subset \{\operatorname{Re} z_1 < 0\} \quad \textit{and} \quad \exists h_1 \in H(U_1),\ h_2 \in H(U_2)$$

with

(6) $$h_1 - h_2 = \frac{\log z_1}{z_1} \textit{ in } U_1 \cap U_2 \quad \textit{and} \quad K \cap U_2 \subset \{\operatorname{Re} z_1 \le 0\}.$$

Then $\exists F$ holomorphic in a neighborhood of K with $F = 1$ on $K \cap \{z_1 = 0\} \cap U_2$ and $|F| < 1$ elsewhere on K.

Proof. By (5) we have in $U_1 \cap U_2$,

$$z_1 h_1 - z_1 h_2 = \log z_1 \quad \text{so} \quad e^{z_1 h_1} = z_1 e^{z_1 h_2}.$$

It follows that if we define

$$f = \begin{cases} e^{z_1 h_1} \text{ in } U_1, \\ z_1 e^{z_1 h_2} \text{ in } U_2, \end{cases}$$

then $f \in H(U_1 \cup U_2)$. Also

(7) $$f \text{ never vanishes on } K \setminus (\{z_1 = 0\} \cap U_2).$$

Assertion. $\exists \varepsilon > 0$ such that if $z \in K \setminus (\{z_1 = 0\} \cap U_2)$, then $f(z)$ lies outside the disk $\{|w - \varepsilon| \le \varepsilon\}$.

Assume first that $z \in U_2$. Then

$$z_1 = e^{-z_1 h_2} f, \qquad \text{so } z_1 h_2 = e^{-z_1 h_2} \cdot h_2 f,$$

or $z_1 h_2 = C \cdot f$, with $C \in H(U_2)$. Hence $z_1 = f e^{-Cf} = f + kf^2$, with $k \in H(U_2)$. By shrinking U_2 we may obtain $|k| \le M$ on U_2, M a constant. Since $\operatorname{Re} z_1 \le 0$ by (6), we have, at z,

$$\begin{aligned} 0 &\ge \operatorname{Re} f + \operatorname{Re}(kf^2) \ge \operatorname{Re} f - |f|^2|k| \\ &\ge \operatorname{Re} f - M|f|^2. \end{aligned}$$

Put $f(z) = w = u + iv$. Then

$$u - M(u^2 + v^2) \leq 0,$$

and so

$$\left(u - \frac{1}{2M}\right)^2 + v^2 \geq \frac{1}{4M^2}.$$

Thus $f(z)$ lies outside the disk:

$$\left|w - \frac{1}{2M}\right| < \frac{1}{2M}.$$

On the other hand, $K \setminus U_2$ is compact and $f \neq 0$ there. Hence for some $r > 0$, $|f(z)| \geq r$ if $z \in K \setminus U_2$. The assertion now follows.

Let D_ε be the disk $\{|w - \varepsilon| \leq \varepsilon\}$ just obtained and put

$$F = -\frac{\varepsilon}{f - \varepsilon}.$$

By choice of D_ε, F is holomorphic in some neighborhood of K. Also on $\{z_1 = 0\} \cap U_2$, $F = 1$ since $f = 0$, and everywhere else on K, $|F| < 1$ since $|f - \varepsilon| > \varepsilon$. Q.E.D.

LEMMA 9.6

Let $\mathfrak{A}$ be a Banach algebra, T a closed subset of $\mathcal{M}$ and U an open neighborhood of T. Suppose that $\exists \phi \in \mathfrak{A}$ with $\hat{\phi} = 1$ on T, $|\hat{\phi}| < 1$ on $U \setminus T$. Then $\exists \Phi \in \mathfrak{A}$ with $\hat{\Phi} = 1$ on T, $|\hat{\Phi}| < 1$ on $\mathcal{M} \setminus T$.

Proof. T and $\mathcal{M} \setminus U$ are disjoint closed subsets of $\mathcal{M}$. Hence $\exists g_2, \ldots, g_n \in \mathfrak{A}$ such that if $\hat{g} : \mathcal{M} \to \mathbf{C}^{n-1}$ is the map $m \to (\hat{g}_2(m), \ldots, \hat{g}_n(m))$, then $\hat{g}(T) \cap \hat{g}(\mathcal{M} \setminus U) = \emptyset$. (Why?)

Put $g_1 = \phi - 1$. Then $\hat{g}_1 = 0$ on T and $\operatorname{Re} \hat{g}_1 < 0$ on $U \setminus T$. Let now $G : \mathcal{M} \to \mathbf{C}^n$ be the map sending $m \to (\hat{g}_1(m), \hat{g}_2(m), \ldots, \hat{g}_n(m))$. The $G(\mathcal{M}) = \sigma(g_1, \ldots, g_n)$. We have

(8) $$G(T) \text{ is a compact subset of } \{z_1 = 0\},$$

(9) $$G(T) \text{ is disjoint from } G(\mathcal{M} \setminus U),$$

(10) $$G(U \setminus T) \subset \{\operatorname{Re} z_1 < 0\}.$$

Choose a neighborhood Δ of $G(T)$ in $\mathbf{C}^n$ with $\bar{\Delta} \cap G(\mathcal{M} \setminus U) = \emptyset$. It is easily seen that $\exists$ an open set D_0 in $\mathbf{C}^n$ such that

(11) $$D_0 \cup \Delta \supset G(\mathcal{M}) \quad \text{and} \quad D_0 \cap \Delta \subset \{\operatorname{Re} z_1 < 0\}.$$

By a construction used in the proof of Theorem 8.2, $\exists C_1, \ldots, C_m \in \mathfrak{A}$ such that if B is the closed subalgebra generated by $g_1, \ldots, g_n, C_1, \ldots, C_m$, then $\sigma_B(g_1, \ldots, g_n) \subset D_0 \cup \Delta$.

Put $\sigma = \sigma(g_1, \ldots, g_n, C_1, \ldots, C_m) \subset \mathbf{C}^{n+m}$, and let $\hat{\sigma}$ be the polynomially convex hull of σ in $\mathbf{C}^{n+m}$. Let π be the natural projection of $\mathbf{C}^{n+m}$ on $\mathbf{C}^n$.

Since $\sigma \subset \sigma_B(g_1, \ldots, g_n, C_1, \ldots, C_m)$, and since the latter set is polynomially convex because $g_1, \ldots, C_m$ generate B, $\hat{\sigma} \subset \sigma_B(g_1, \ldots, g_n, C_1, \ldots, C_m)$, and so

$$\pi(\hat{\sigma}) \subset \pi(\sigma_B(g_1, \ldots, C_m)) = \sigma_B(g_1, \ldots, g_n).$$

Thus $\pi(\hat{\sigma}) \subset D_0 \cup \Delta$, and so

$$\hat{\sigma} \subset \pi^{-1}(D_0) \cup \pi^{-1}(\Delta). \tag{12}$$

Because of (11) we have

$$\pi^{-1}(D_0) \cap \pi^{-1}(\Delta) \subset \{\operatorname{Re} z_1 < 0\}. \tag{13}$$

Now $\hat{\sigma}$ is polynomially convex and $(\log z_1)/z_1$ is holomorphic in $\pi^{-1}(D_0) \cap \pi^{-1}(\Delta)$. Lemma 9.4 then yields a neighborhood W of $\hat{\sigma}$, and $h_1 \in H(\pi^{-1}(D_0) \cap W)$, $h_2 \in H(\pi^{-1}(\Delta) \cap W)$ such that

$$h_1 - h_2 = \frac{\log z_1}{z_1} \qquad \text{in } \pi^{-1}(D_0) \cap \pi^{-1}(\Delta) \cap W.$$

We now apply Lemma 9.5 with $\sigma = K$, $U_1 = \pi^{-1}(D_0) \cap W$, and $U_2 = \pi^{-1}(\Delta) \cap W$. Since $\sigma \subseteq \hat{\sigma}$, hypotheses (4) and (5) hold. By choice of Δ and (10), $G(\mathscr{M}) \cap \Delta \subset \{\operatorname{Re} z_1 \leq 0\}$, whence $\sigma \cap \pi^{-1}(\Delta) \subset \{\operatorname{Re} z_1 \leq 0\}$. So hypothesis (6) also holds. We conclude the existence of F holomorphic in a neighborhood of σ with $F = 1$ on $\{z_1 = 0\} \cap \pi^{-1}(\Delta) \cap \sigma$ and $|F| < 1$ elsewhere on σ.

By Theorem 8.2, $\exists \Phi \in \mathfrak{A}$ with

$$\hat{\Phi}(M) = F(\hat{g}_1(M), \ldots, \hat{g}_n(M), \hat{C}_1(M), \ldots, \hat{C}_m(M))$$

for all $M \in \mathscr{M}$. For $M \in T$, the corresponding point of σ is in $\{z_1 = 0\} \cap \pi^{-1}(\Delta)$, so $\hat{\Phi}(M) = 1$. For $M \in \mathscr{M} \setminus T$, the corresponding point of σ is not in $\{z_1 = 0\} \cap \pi^{-1}(\Delta)$, so $|\hat{\Phi}(M)| < 1$.

Q.E.D.

Proof of Theorem 9.3. Suppose that (3) is false. Choose $x_0 \in \bar{U}$ with $|\hat{f}(x_0)| = \max_{\bar{U}} |\hat{f}|$. Then

$$|\hat{f}(x_0)| > \max_{\partial U} |\hat{f}|. \tag{14}$$

Without loss of generality, $\hat{f}(x_0) = 1$. Let $T = \{y \in \bar{U} | \hat{f}(y) = 1\}$. Then T is compact and $\subset U$. Put $\phi = \frac{1}{2}(1 + f)$. Then $\phi \in \mathfrak{A}$, $\hat{\phi} = 1$ on T, $|\hat{\phi}| < 1$ on $U \setminus T$.

Lemma 9.6 now supplies $\Phi \in \mathfrak{A}$, with $\hat{\Phi} = 1$ on T, $|\hat{\Phi}| < 1$ on $\mathscr{M} \setminus T$. Since $U \subset \mathscr{M} \setminus \check{S}(\mathfrak{A})$, we get that $|\hat{\Phi}| < 1$ on $\check{S}(\mathfrak{A})$. This is impossible, and so (3) holds. Q.E.D.

Note. Some, but not all, of the following exercises depend on Theorem 9.3.

Exercise 9.6. Let $\mathfrak{A}$ be a Banach algebra and assume that $\check{S}(\mathfrak{A}) \neq \mathscr{M}$. Show that the restriction of $\hat{\mathfrak{A}}$ to $\check{S}(\mathfrak{A})$ is not uniformly dense in $C(\check{S}(\mathfrak{A}))$.

Exercise 9.7. Let $\mathfrak{A}$ be a Banach algebra and assume that $\check{S}(\mathfrak{A}) \neq \mathscr{M}$. Show that $\check{S}(\mathfrak{A})$ is uncountable.

Exercise 9.8. Let $\mathfrak{A}$ be a Banach algebra and fix $p \in \check{S}(\mathfrak{A})$. Assume that p is an isolated point of $\check{S}(\mathfrak{A})$, viewed in the topology induced on $\check{S}(\mathfrak{A})$ by $\mathscr{M}$. Show that p is then an isolated point of $\mathscr{M}$.

THEOREM 9.7

Let $\mathfrak{A}$ be a uniform algebra on a space X. Let $U_1, U_2, \ldots, U_s$ be an open covering of $\mathcal{M}$. Denote by $\mathcal{L}$ the set of all f in $C(\mathcal{M})$ such that for $j = 1, \ldots, s$, $f|_{U_j}$ lies in the uniform closure of $\hat{\mathfrak{A}}|_{U_j}$. Then $\mathcal{L}$ is a closed subalgebra of $C(\mathcal{M})$ and $\check{S}(\mathcal{L}) \subseteq X$.

Proof. The proof is a corollary of Theorem 9.3. We leave it to the reader as *Exercise 9.9.

Exercise 9.10. Is Theorem 9.3 still true if we omit the assumption $U \subset \mathcal{M} \setminus \check{S}(\mathfrak{A})$?

NOTES

Theorem 9.1 is due to G. E. Šilov, On the extension of maximal ideals, *Dokl. Acad. Sci. URSS* (N.S.) (1940), 83–84. The proof given here, which involves no transfinite induction or equivalent argument, is due to Hörmander [8, Theorem 3.1.18]. Theorem 9.3 is due to H. Rossi, The local maximum modulus principle, *Ann. Math.* **72**, No. 1 (1960), 1–11. The proof given here is in the book by Gunning and Rossi [6, pp. 62–63].

10

MAXIMALITY AND RADÓ'S THEOREM

Let X be a compact space and $\mathfrak{A}$ a uniform algebra on X. Denote by $\|\ \|$ the uniform norm on $C(X)$. Note that if $x,y \in \mathfrak{A}$, then $x + \bar{y} \in C(X)$, so that $\|x + \bar{y}\|$ is defined.

LEMMA 10.1 (PAUL COHEN)

Let $a,b \in \mathfrak{A}$. Assume that

$$\|1 + a + \bar{b}\| < 1.$$

Then $a + b$ is invertible in $\mathfrak{A}$.

Note. When $b = 0$, this of course holds in an arbitrary Banach algebra.

Proof. Put $f = a + b$. We have

$$\|1 + a + \bar{b}\| < 1, \qquad \text{hence } \|1 + \bar{a} + b\| < 1,$$

whence

$$\|1 + a + \bar{b} + 1 + \bar{a} + b\| < 2 \qquad \text{or} \qquad k = \|1 + \operatorname{Re} f\| < 1.$$

For all $x \in X$, then

$$|1 + \operatorname{Re} f(x)| \leq k.$$

This means that $f(x)$ lies in the left-half plane for all x, which suggests that for small $\varepsilon > 0$,

$$1 + \varepsilon f(x)$$

lies in the unit disk for all x. Indeed,

$$\begin{aligned} |1 + \varepsilon f(x)|^2 &= 1 + \varepsilon^2 |f(x)|^2 + 2\varepsilon \operatorname{Re} f(x) \\ &\leq 1 + c\varepsilon^2 + 2d\varepsilon, \end{aligned}$$

where $c = \|f\|^2$ and $d = -1 + k < 0$. Hence for small $\varepsilon > 0$, $|1 + \varepsilon f(x)| < 1$ for all x, or $\|1 + \varepsilon f\| < 1$, as we had guessed.

It follows that εf is invertible in $\mathfrak{A}$ for some ε and so f is invertible. Q.E.D.

We shall now apply this lemma to a particular algebra. Let D = closed unit disk in the z-plane and Γ the unit circle. Let $A(D)$ be the space of all functions analytic in $\mathring{D}$ and continuous in D. Put

$$\mathfrak{A}_0 = A(D)|_\Gamma$$

and give $\mathfrak{A}_0$ the uniform norm on Γ. $\mathfrak{A}_0$ is then isomorphic and isometric to $A(D)$ and is a uniform algebra on Γ. The elements of $\mathfrak{A}_0$ are precisely those functions in $C(\Gamma)$ that admit an analytic extension to $|z| < 1$.

$\mathfrak{A}_0$ is approximately one half of $C(\Gamma)$. For the functions

$$e^{in\theta}, \qquad n = 0, \pm 1, \pm 2, \ldots$$

span a dense subspace of $C(\Gamma)$, while $\mathfrak{A}_0$ contains exactly those $e^{in\theta}$ with $n \geq 0$.

Exercise 10.1. Put $g = \Sigma_{-p}^{p} c_\nu e^{i\nu\theta}$, where the c_ν are complex constants. Compute the closed algebra generated by $\mathfrak{A}_0$ and g, i.e., the closure in $C(\Gamma)$ of all sums

$$\sum_{\nu=0}^{N} a_\nu g^\nu, \qquad a_\nu \in \mathfrak{A}_0.$$

THEOREM 10.2 (MAXIMALITY OF $\mathfrak{A}_0$)

Let B be a uniform algebra on Γ with

$$\mathfrak{A}_0 \subseteq B \subseteq C(\Gamma).$$

Then either $\mathfrak{A}_0 = B$ or $B = C(\Gamma)$.

We shall deduce this result by means of Lemma 10.1 as follows. Assuming $B \neq \mathfrak{A}_0$, we construct elements $u,v \in B$ with

$$\|1 + z \cdot u + \bar{z}\bar{v}\| < 1, \tag{1}$$

where $z = e^{i\theta}$. Then we conclude that $zu + zv$ is invertible in B, when z is invertible in B. Hence $B \supset e^{in\theta}$, $n = 0, \pm 1, \pm 2, \ldots$, so $B = C(\Gamma)$, as required. To construct u and v we argue as follows: For each $h \in C(\Gamma)$, put

$$h_k = \frac{1}{2\pi}\int_0^{2\pi} h(e^{i\theta})e^{-ik\theta}\,d\theta, \qquad k = 0, \pm 1, \pm 2, \ldots.$$

Exercise 10.2. Let $h \in C(\Gamma)$. Prove that $h \in \mathfrak{A}_0$ if and only if $h_k = 0$, for all $k < 0$.

Suppose now that $B \neq \mathfrak{A}_0$. Hence $g \in B$ with $g_k \neq 0$, for some $k < 0$. Without loss of generality we may suppose that $g_{-1} = 1$. (Why?)

Choose a trigonometric polynomial T with

$$\|g - T\| < 1. \tag{2}$$

We can assume $T_{-1} = 1$, or

$$T = \sum_{-N}^{-2} T_\nu z^\nu + z^{-1} + \sum_{0}^{N} T_\nu z^\nu.$$

Hence

$$zT = \sum_{-N}^{-2} T_\nu z^{\nu+1} + 1 + z\sum_{0}^{N} T_\nu z^\nu$$
$$= \bar{z}\cdot\bar{P} + 1 + zQ,$$

where P and Q are polynomials in z. Equation (2) gives

$$\|zg - zT\| < 1 \qquad \text{or} \qquad \|z(Q-g) + \bar{z}\bar{P} + 1\| < 1.$$

Also, $Q - g \in B$, $P \in B$, so we have (1), and we are done.

THEOREM 10.3 (RUDIN)

Let $\mathscr{L}$ be an algebra of continuous functions on D such that
(a) *The function z is in $\mathscr{L}$.*
(b) *$\mathscr{L}$ satisfies a maximum principle relative to Γ:*

$$|G(x)| \le \max_{\Gamma}|G|, \qquad \text{all } x \in D,\ G \in \mathscr{L}.$$

Then $\mathscr{L} \subseteq A(D)$.

Proof. The uniform closure of $\mathscr{L}$ on D, written $\mathfrak{A}$, still satisfies (a) and (b).

Put $B = \mathfrak{A}|_\Gamma$. Because of (b), B is closed under uniform convergence on Γ and by (a), $\mathfrak{A}_0 \subseteq B$. So Theorem 10.2 applies to yield $B = \mathfrak{A}_0$ or $B = C(\Gamma)$.

Consider the map $g \to G(0)$ for $g \in B$, where G is the function in $\mathfrak{A}$ with $G = g$ on Γ. By (b), G is unique. The map is a homomorphism of $B \to \mathbf{C}$ and is not evaluation at a point of Γ. (Why?) Hence $B \neq C(\Gamma)$, and so $B = \mathfrak{A}_0$.

Fix $F \in \mathfrak{A}$. $F|_\Gamma \in \mathfrak{A}_0$, so $\exists F^* \in A(D)$ with $F = F^*$ on Γ. $F - F^*$ then $\in \mathfrak{A}$ and by (b) vanishes identically on D. So $F \in A(D)$ and thus $\mathfrak{A} = A(D)$, whence the assertion.

Now let X be any compact space, $\mathscr{L}$ an algebra of continuous functions on X, and X_0 a boundary for $\mathscr{L}$ in the sense of Definition 9.1; i.e., X_0 is a closed subset of X with

(3) $$|g(x)| \le \max_{X_0}|g|, \qquad \text{all } g \in \mathscr{L},\ x \in X.$$

LEMMA 10.4 (GLICKSBERG)

Let E be a subset of X_0 and let $f \in \mathscr{L}$ and $f = 0$ on E. Then for each $x \in X$ either
(a) $f(x) = 0$, *or*
(b) $|g(x)| \le \sup_{X_0 \setminus E}|g|$, *all* $g \in \mathscr{L}$.

Proof. Fix $g \in \mathscr{L}$. Then $f \cdot g \in \mathscr{L}$. Fix $x \in X$ with $f(x) \neq 0$. We have

$$|(fg)(x)| \le \max_{X_0}|fg| = \sup_{X_0 \setminus E}|fg|$$
$$\le \sup_{X_0\setminus E}|f| \cdot \sup_{X_0\setminus E}|g|.$$

Hence

$$|g(x)| \leq K \sup_{X_0 \setminus E} |g|,$$

where $K = |f(x)|^{-1} \cdot \sup_{X_0 \setminus E}|f|$. Applying this to g^n, $n = 1, 2, \ldots$ gives

$$|g(x)|^n = |g^n(x)| \leq K \sup_{X_0 \setminus E} |g^n| = K(\sup_{X_0 \setminus E} |g|)^n.$$

Taking nth roots and letting $n \to \infty$ gives (b). Q.E.D.

Consider now the following classical result: Let Ω be a bounded plane region and z_0 a nonisolated boundary point of Ω. Let U be a neighborhood of z_0 in $\mathbf{C}$.

THEOREM 10.5

Let $f \in A(\Omega)$ and assume that $f = 0$ on $\partial\Omega \cap U$. Then $f \equiv 0$ in Ω.

If we assume that

(4) $$\exists \text{ a sequence } \{z_n\} \text{ in } \mathbf{C} \setminus \bar{\Omega} \text{ with } z_n \to z_0,$$

then Lemma 10.4 gives a direct proof, as follows.

Put $X = \bar{\Omega}$, $\mathscr{L} = A(\Omega)$. Then $\partial\Omega$ is a boundary for $\mathscr{L}$. Put $E = \partial\Omega \cap U$.

With z_n as in (4), put

$$g_n(z) = \frac{1}{z - z_n}.$$

Then $g_n \in \mathscr{L}$. If $\varepsilon > 0$ is small enough, we have for all $x \in \Omega$ with $|x - z_0| < \varepsilon$,

$$|g_n(x)| > \sup_{\partial\Omega \setminus E} |g_n|$$

for all large n. Hence the lemma gives $f(x) = 0$ for all $x \in \Omega$ with $|x - z_0| < \varepsilon$, and so $f \equiv 0$. Q.E.D.

If we do not assume (4), the conclusion follows from

THEOREM 10.6 (RADÓ'S THEOREM)

Let h be a continuous function on the disk D. Let Z denote the set of zeros of h. If h is analytic on $\mathring{D} \setminus Z$, then h is analytic on $\mathring{D}$.

Proof. We assume that Z has an empty interior. The case $\mathring{Z} \neq \emptyset$ is treated similarly.

Let $\mathscr{L}$ consist of all sums

$$\sum_{\nu=0}^{N} a_\nu h^\nu, \qquad a_\nu \in A(D).$$

If $f \in \mathscr{L}$, f is analytic in $|z| < 1$ except possibly on Z, so

(5) $$|f(x)| \leq \max_{\Gamma \cup Z} |f|, \qquad \text{all } x \in D.$$

We apply Lemma 10.4 to $\mathscr{L}$ with $X_0 = \Gamma \cup Z$, $E = Z$. Since $h \in \mathscr{L}$ and $h = 0$ on Z we get by the lemma

(6) $$|g(x)| \leq \sup_{\Gamma}|g|, \qquad \text{all } g \in \mathscr{L},$$

if $x \in D \setminus Z$, since then $h(x) \neq 0$.

By continuity, (6) then holds for all $x \in D$. Thus $\mathscr{L}$ satisfies the hypotheses of Theorem 10.3, and so $\mathscr{L} \subseteq A(D)$. Thus h is analytic on $\mathring{D}$. Q.E.D.

Note that Theorem 10.5 follows at once from Radó's theorem.

For future use we next prove

THEOREM 10.7

Let $\mathfrak{A}$ be a uniform algebra on a space X with maximal ideal space $\mathscr{M}$. Let $f \in \mathfrak{A}$ satisfying

(a) $|f| = 1$ on X.

(b) $0 \in f(\mathscr{M})$.

(c) $\exists$ a closed subset Γ_0 of Γ having positive linear measure such that for each $\lambda \in \Gamma_0$ there is a unique point q in X with $f(q) = \lambda$.

Then

(7) *For each $z_1 \in \mathring{D}$ there is a unique x in $\mathscr{M}$ with $f(x) = z_1$.*

(8) *If $g \in \mathfrak{A}$, $\exists G$ analytic in $\mathring{D}$ such that*

$$g = G(f) \qquad \text{on } f^{-1}(\mathring{D}).$$

Proof. For each measure μ on X, let $f(\mu)$ denote the induced measure on Γ; i.e., for $S \subset \Gamma$,

$$f(\mu)(S) = \mu(f^{-1}(S)).$$

where $f^{-1}(S) = \{x \in X | f(x) \in S\}$.

Since by (b), $f(\mathscr{M})$ contains 0, and by (a), $f(X) \subset \Gamma$, it follows that $f(\mathscr{M}) \supset D$. (Why? See Lemma 11.1.) Fix p_1 and p_2 in $\mathscr{M}$ with

$$f(p_1) = f(p_2) = z_1 \in \mathring{D}.$$

We must show that $p_1 = p_2$. Suppose not. Then $\exists g \in \mathfrak{A}$ with $g(p_1) = 1$ and $g(p_2) = 0$. Choose, by Exercise 1.2, positive measures μ_1 and μ_2 on X with

$$h(p_j) = \int h \, d\mu_j, \qquad \text{all } h \in \mathfrak{A},$$

for $j = 1, 2$.

Let G be a polynomial. Then

$$\int G \, d(f(\mu_1)) = \int G(f) \, d\mu_1 = G(f(p_1))$$

and similarly for μ_2. Hence $f(\mu_1) - f(\mu_2)$ is a real measure on Γ annihilating the polynomials. Hence $f(\mu_1) - f(\mu_2) = 0$. (Why?)

Since by (c), f maps $f^{-1}(\Gamma_0)$ bijectively on Γ_0, it follows that μ_1 and μ_2 coincide when restricted to $f^{-1}(\Gamma_0)$. Hence the same holds for the measure $g\mu_1$ and $g\mu_2$.

Put $\lambda_j = f(g\mu_j)$, $j = 1, 2$. Then λ_1 and λ_2 coincide when restricted to Γ_0. For a polynomial G we have

$$\int G\,d\lambda_j = \int G(f)g\,d\mu_j = G(f(p_j))g(p_j).$$

Hence by choice of g,

$$\int G\,d\lambda_1 = G(z_1),$$

$$\int G\,d\lambda_2 = 0.$$

Thus

$$\int G\,d(\lambda_1 - \lambda_2) = G(z_1), \qquad \text{all } G. \tag{9}$$

It follows that the measure $(z - z_1)\,d(\lambda_1 - \lambda_2)$ is orthogonal to all polynomials. By the theorem of F. and M. Riesz (see [7, Chap. 4]), $\exists k \in H^1$ with

$$(z - z_1)\,d(\lambda_1 - \lambda_2) = k\,dz.$$

It follows that $k = 0$ on Γ_0. Since Γ_0 has positive measure, $k \equiv 0$. (See [7, Chap. 4].) But $z - z_1 \neq 0$ on Γ, so $\lambda_1 - \lambda_2 = 0$, contradicting (9). Hence $p_1 = p_2$, and (7) is proved.

It follows from (7) that if $g \in \mathfrak{A}$, $\exists G$ continuous on $\mathring{D}$, with $g = G(f)$ on $f^{-1}(\mathring{D})$. It remains to show that G is analytic.

Fix an open disk U with closure $\overline{U} \subset \mathring{D}$. Let $\mathscr{L}$ be the algebra of all functions

$$G = g(f^{-1}), \qquad g \in \mathfrak{A},$$

restricted to $\overline{U}$.

Choose $x \in U$. $f^{-1}(U)$ is an open subset of $\mathscr{M}$ with boundary $f^{-1}(\partial U)$, and $f^{-1}(x) \in f^{-1}(U)$.

By the local maximum modulus principle, if $h \in \mathfrak{A}$,

$$|h(f^{-1}(x))| \leq \max_{f^{-1}(\partial U)} |h|$$

or

$$|H(x)| \leq \max_{\partial U} |H|$$

if $H = h(f^{-1}) \in \mathscr{L}$. Note also that $z = f(f^{-1}) \in \mathscr{L}$.

Theorem 10.3 (which clearly holds if D is replaced by an arbitrary disk) now applies to the algebra $\mathscr{L}$ on $\overline{U}$. We conclude that $\mathscr{L} \subseteq A(\overline{U})$, and so $G = g(f^{-1})$ is analytic in U for every $g \in \mathfrak{A}$.

Thus G is analytic in $\mathring{D}$, whence (8) holds. Q.E.D.

NOTES

Lemma 10.1 and the proof of Theorem 10.2 based on it are due to Paul Cohen, A note on constructive methods in Banach algebras, *Proc. Am. Math. Soc.* **12** (1961). Theorem 10.2 is due to the author, On algebras of continuous functions, *Proc. Am. Math. Soc.* **4** (1953). Paul Cohen's proof of Theorem 10.2 developed out of an abstract proof of the same result by K. Hoffman and I. M. Singer, Maximal algebras of continuous functions, *Acta Math.* **103** (1960). Theorem 10.3 is due to W. Rudin, Analyticity and the maximum modulus principle, *Duke Math. J.* **20** (1953). Lemma 10.4 is a result of I. Glicksberg, Maximal algebras and a theorem of Radó, *Pacific J. Math.* **14** (1964). Theorem 10.6 is due to T. Radó and has been given many proofs. See, in particular, E. Heinz, Ein elementarer Beweis des Satzes von Radó–Behnke–Stein–Cartan. The proof we have given is to be found in the paper of Glicksberg cited above. Theorem 10.7 is due to E. Bishop and is contained in Lemma 13 of his paper [2].

11

ANALYTIC STRUCTURE

Let D denote the closed unit disk and consider the algebra $A(D)$.

Exercise 11.1. The maximal ideal space of $A(D)$ is naturally identified with D, and the Šilov boundary of $A(D)$ is ∂D.

Fix f in $A(D)$. $f(\partial D)$ is a certain compact set. Let W be a component of $\mathbf{C} \setminus f(\partial D)$. The following facts are classical:

(a) If f takes some value in W at a point of D, then f takes on every value in W.

(b) If $\lambda_0 \in W$, $\{z \in D | f(z) = \lambda_0\}$ is finite.

Now let $\mathfrak{A}$ be an arbitrary Banach algebra, $\mathcal{M}$ its maximal ideal space, and X its Šilov boundary. How can we generalize properties (a) and (b)? The first is easy.

LEMMA 11.1

Let $f \in \mathfrak{A}$ and let W be a component of $\mathbf{C} \setminus f(X)$. Fix $\lambda_0 \in W$. If f takes the value λ_0 on $\mathcal{M}$, then it takes on every value in W on $\mathcal{M}$.

Proof. Let

(1) $$W_1 = \{\lambda \in W | (f - \lambda)^{-1} \in \mathfrak{A}\}.$$

W_1 is open. (Why?) Also, W_1 is a closed subset of W. For let $\lambda^* \in W$ and let $\lambda_n \in W_1$ and $\lambda_n \to \lambda^*$ as $n \to \infty$. Suppose that $\lambda^* \notin W_1$. It follows that $\exists p^* \in \mathcal{M}$ with $f(p^*) = \lambda^*$. (Why?) Also $(f - \lambda_n)^{-1} \in \mathfrak{A}$. Then

$$\max_X |(f - \lambda_n)^{-1}| \geq |(f - \lambda_n)^{-1}(p^*)|$$

$$= \left| \frac{1}{\lambda^* - \lambda_n} \right| \to \infty \qquad \text{as } n \to \infty.$$

But $\lambda_n \to \lambda^*$ and $\lambda^* \notin f(X)$, and so

(2) $$\max_X |(f - \lambda_n)^{-1}|$$

is bounded as $n \to \infty$. This contradiction shows that $\lambda^* \in W_1$.

Thus W_1 is an open and closed subset of W. Also $\lambda_0 \notin W_1$. Hence W_1 is empty and so the assertion of the lemma holds.

On the other hand, condition (b) does not hold in general. (Think of an example.) In fact, condition (b) has extremely strong implications for the structure of $\mathscr{M}$.

Definition 11.1. Let $\mathfrak{A}$ be a Banach algebra, $\mathscr{M}$ its maximal ideal space, and $p \in \mathscr{M}$. Let Φ be a one-to-one continuous map from $|z| < 1$ into $\mathscr{M}$ with $\Phi(0) = p$ such that, for every $h \in \mathfrak{A}$, $h \circ \Phi$ is analytic in $|z| < 1$. We then call the set $\{\Phi(z) \,|\, |z| < 1\}$ an *analytic disk through p.*

The main result of this section is the following:

THEOREM 11.2

Let $\mathfrak{A}$ be a uniform algebra on a space X and $\mathscr{M}$ its maximal ideal space. Fix $f \in \mathfrak{A}$ and let W be a component of $\mathbf{C} \setminus f(X)$. Assume that there exists a set of positive plane measure $G \subset W$ so that

$$f^{-1}(\lambda) = \{p \in \mathscr{M} \,|\, f(p) = \lambda\}$$

is a finite set for each $\lambda \in G$.

Then each point p in

$$f^{-1}(W) = \{q \in \mathscr{M} \,|\, f(q) \in W\}$$

has a neighborhood in $\mathscr{M}$ which is a finite union of analytic disks through p.

Note. If S is a subset of W, $f^{-1}(S)$ will denote $\{p \in \mathscr{M} \,|\, f(p) \in S\}$.

Proof. For each j, put

$$W_j = \{z \in W \,|\, f^{-1}(z) \text{ has } j \text{ elements}\}.$$

$G \subset \bigcup_{j=1}^{\infty} W_j$. Since G has positive measure, there is some k such that W_k has positive measure. Fix such a k.

Choose z_0 in W_k so that z_0 is a point of density of W_k. We have

$$f^{-1}(z_0) = \{p_1, p_2, \ldots, p_k\}.$$

Let $U_1, \ldots, U_k$ be disjoint neighborhoods of $p_1, \ldots, p_k$ in $\mathscr{M}$. Choose a closed disk D centered at z_0 such that

$$f^{-1}(D) \subset \bigcup_{\nu=1}^{k} U_\nu.$$

Put B = boundary of D. Since z_0 is a point of density of W_k, we may suppose, by Fubini's theorem, that there is a closed set B_0 having linear measure > 0 with $B_0 \subseteq B \cap W_k$.

Without loss of generality, we may assume that $z_0 = 0$ and D is the unit disk.

For each closed set $K \subset \mathcal{M}$, denote by $\mathfrak{A}(K)$ the uniform closure on K of the restriction of $\mathfrak{A}$ to K. $\mathfrak{A}(K)$ is a uniform algebra on K and K has a natural embedding in $\mathcal{M}(\mathfrak{A}(K))$.

(3) $$\mathcal{M}(\mathfrak{A}(f^{-1}(D)) = f^{-1}(D). \qquad \text{(Why?)}$$

The topological boundary of $f^{-1}(D)$ is contained in $f^{-1}(B)$. Let $p \in f^{-1}(D)$. By the local maximum modulus principle we get

$$|g(p)| \leq \max_{f^{-1}(B)} |g|, \qquad \text{all } g \in \mathfrak{A}.$$

Hence we have

(4) $$\check{S}(\mathfrak{A}(f^{-1}(D))) \subseteq f^{-1}(B).$$

Let J be a connected component of $f^{-1}(D)$. We claim that

(5) $$\mathcal{M}(\mathfrak{A}(J)) = J.$$

Let m be a homomorphism of $\mathfrak{A}(J) \to \mathbf{C}$. Denote by $\tilde{m}$ the homomorphism induced by m on $\mathfrak{A}$. $\tilde{m} \in \mathcal{M}$. In fact, $\tilde{m} \in f^{-1}(D)$. Suppose that $\tilde{m} \notin J$. Since J is a connected component of $f^{-1}(D)$, we can choose an open and closed subset J' of $f^{-1}(D)$ with $J \subset J'$ and $\tilde{m} \notin J'$. Then $\exists e \in \mathfrak{A}(f^{-1}(D))$ with $e(\tilde{m}) = 1$ and $e = 0$ on J'. (Why?) Hence we can find $e' \in \mathfrak{A}$ with e' nearly 1 at $\tilde{m}$ and e' nearly 0 on J'. Thus, regarding e' as an element of $\mathfrak{A}(J)$,

$$|m(e')| = |e'(\tilde{m})| > \max_J |e'| = \|e'\|_{\mathfrak{A}(J)}.$$

This contradicts the fact that m has norm 1 as linear functional on $\mathfrak{A}(J)$. So $\tilde{m} \in J$. This yields (5). We next claim that

(6) $$\check{S}(\mathfrak{A}(J)) \subseteq f^{-1}(B) \cap J.$$

Otherwise $\check{S}(\mathfrak{A}(J))$ meets $J \setminus f^{-1}(B)$. Then $\exists y$ in $J \setminus f^{-1}(B)$ and $g_0 \in \mathfrak{A}(J)$ with

$$|g_0(y)| > \max_{J \cap f^{-1}(B)} |g_0|.$$

It follows that we can take $g \in \mathfrak{A}$ with the same property.

We can choose a set Ω open and closed in $f^{-1}(D)$, containing J, and lying in a prescribed neighborhood of J. In particular, we can choose Ω so that

$$|g(y)| > \max_{\Omega \cap f^{-1}(B)} |g|.$$

Since Ω is open and closed, we can find $e \in \mathfrak{A}(f^{-1}(D))$ with $e = 1$ on Ω and $e = 0$ on $f^{-1}(D) \setminus \Omega$. Then

$$|(eg)(y)| > \max_{\Omega \cap f^{-1}(B)} |eg|.$$

Also $eg = 0$ on $f^{-1}(B) \setminus \Omega$. Thus

$$|(eg)(y)| > \max_{f^{-1}(B)} |eg|.$$

But $eg \in \mathfrak{A}(f^{-1}(D))$ and so the last inequality contradicts (4). Thus (6) holds.

(7) Let J be a connected component of $f^{-1}(D)$ such that $f(J)$ meets $\mathring{D}$. Then $f(J) \supset D$.

For by (6) f maps $\check{S}(\mathfrak{A}(J))$ into B. Since $f(J)$ meets $\mathring{D}$, Lemma 11.1 applied to $\mathfrak{A}(J)$ gives that f maps J onto $\mathring{D}$. Since $f(J)$ is closed, (7) follows.

Let $p_1, \ldots, p_k$ be, as above, the points f maps on 0. For each ν, let J_ν be the component of $f^{-1}(D)$ which contains p_ν. By the choice of D, the J_ν are disjoint closed subsets of $\mathcal{M}$. By (7), $f(J_\nu) \supset D$ for each ν.

Each point in $W_k \cap D$ is hence taken on exactly once by f in each J_ν. Recall the subset B_0 of B introduced above. Fix $z \in B_0$.

Fix ν. $\exists$ unique point q in J_ν with $f(q) = z$. Since $z \in B$, it follows that $q \in \check{S}(\mathfrak{A}(J_\nu))$. (Why?)

The hypotheses of Theorem 10.7 are thus satisfied by $\mathfrak{A}(J_\nu)$, viewed as a uniform algebra on $\check{S}(\mathfrak{A}(J_\nu))$.

This theorem now yields

Assertion 1. Fix ν, $1 \leq \nu \leq k$. Then f maps $J_\nu \cap f^{-1}(\mathring{D})$ in a one-to-one fashion onto $\mathring{D}$, and every $g \in \mathfrak{A}$ admits on $J_\nu \cap f^{-1}(\mathring{D})$ a representation

$$g = G(f)$$

with G analytic on $\mathring{D}$.

We next claim that

$$f^{-1}(\mathring{D}) \subseteq \bigcup_{\nu=1}^{k} J_\nu. \tag{8}$$

For suppose that $\exists x \in f^{-1}(\mathring{D})$ with $x \notin \bigcup_\nu J_\nu$. Let K be the connected component of $f^{-1}(D)$ through x. For each ν, K is disjoint from J_ν. Also, by (7), $0 \in f(K)$. Thus $f^{-1}(0)$ contains more than k points, which is a contradiction. So (8) holds.

Now fix a compact subdisk Δ of $\mathring{D}$, centered at 0, and fix ν. Put

$$U_\nu = f^{-1}(\mathring{\Delta}) \cap J_\nu.$$

Because of (8), U_ν is an open subset of $\mathcal{M}$. Put

$$\mathcal{M}_1 = \mathcal{M} \setminus \bigcup_{\nu=1}^{k} U_\nu.$$

Then $\mathcal{M}_1$ is compact and $f \neq 0$ on $\mathcal{M}_1$.

Each $x \in \mathcal{M}_1$ has a neighborhood U_x such that $1/f$ is uniformly approximable on U_x by polynomials in f. Let $U_{x_1}, \ldots, U_{x_l}$ be a finite subset covering $\mathcal{M}_1$. Then, putting $U_{k+j} = U_{x_j}$,

$$U_1, \ldots, U_k, U_{k+1}, \ldots, U_{k+l}$$

together form an open covering of $\mathcal{M}$.

For each i, denote by $\mathfrak{A}(U_i)$ the uniform closure on U_i of the restriction of $\mathfrak{A}$ to U_i. We introduce the algebra $\mathcal{L}$ consisting of all g in $C(\mathcal{M})$ such that $g|U_i \in \mathfrak{A}(U_i)$ for each i, $1 \leq i \leq k + l$.

Assertion 2. Let $F \in \mathscr{L}$ and assume that $F(p_j) = 0$, $1 \leq j \leq k$. Then F/f, appropriately defined at the p_j, belongs to $\mathscr{L}$.

If $i \leq k$, we shall now show that $F/f \in \mathfrak{A}(U_i)$.

Since $F \in \mathscr{L}$, $\exists F_n \in \mathfrak{A}$ with $F_n \to F$ uniformly on U_i. By Assertion 1, $\exists \Phi_n$ analytic on $\mathring{\Delta}$, continuous on Δ, with

$$F_n = \Phi_n(f) \text{ on } U_i.$$

Since F_n converges uniformly on U_i, and hence on $\overline{U}_i$, Φ_n converges uniformly on Δ to some function Φ. Then Φ is analytic on $\mathring{\Delta}$ and continuous on Δ. We have

$$F = \Phi(f) \text{ on } U_i.$$

Also, $\Phi(0) = 0$. We can find a sequence P_j of polynomials with $P_j(z) \to \Phi(z)/z$ uniformly on Δ. Hence $P_j(f)$ converges uniformly on U_i to a continuous function which equals $\Phi(f)/f = F/f$ on $U_i \setminus \{p_i\}$. So $F/f \in \mathfrak{A}(U_i)$, as desired.

For $i = k + 1, \ldots, k + l$, $1/f$ is uniformly approximable on U_i by polynomials in f. Hence again F/f is in $\mathfrak{A}(U_i)$, as desired.

Thus Assertion 2 is proved. It follows that the quotient algebra $\mathscr{L}/(f)$, where (f) is the ideal in $\mathscr{L}$ generated by f, has dimension k. (Why?)

Further, by Theorem 9.7, $\check{S}(\mathscr{L}) \subset X$. Hence $f(\check{S}(\mathscr{L})) \subseteq f(X)$, and so $W \subseteq \mathbb{C} \setminus f(\check{S}(\mathscr{L}))$.

Assertion 3. For every $\lambda \in W$, there are at most k points in $f^{-1}(\lambda)$.

We appeal to

Exercise 11.2. *Let T be a continuous transformation of a Banach space Y into itself. Assume that*

(a) *T is one-to-one.*

(b) *The range of T is closed.*

(c) *The range of T has codimension α, where α may be finite or ∞.*

Then for all continuous linear transformation T' with $\|T' - T\|$ sufficiently small, the range of T' has codimension α.

General results containing the assertion of Exercise 11.2 are proved in Chapter 4 of the book by T. Kato, *Perturbation Theory for Linear Operators*, Springer-Verlag, Berlin, 1966.

We shall apply this to the Banach space $\mathscr{L}$ with $\|\phi\| = \max_{\check{S}(\mathscr{L})} |\phi|$. For each $\lambda \in W$ we define the linear transformation

$$T_\lambda : g \to (f - \lambda) \cdot g$$

of $\mathscr{L} \to \mathscr{L}$.

Fix $\lambda \in W$. Then T_λ is one-to-one and has closed range. (Why?) Put $\beta(\lambda) =$ codimension of the range of T_λ. Exercise 11.2 gives that β is locally constant on W. Hence β is constant on W. But $\beta(0) = k$. Thus we get that $\mathscr{L}/(f - \lambda)$ has dimension k for each $\lambda \in W$.

It follows that if $\lambda \in W$, then there exist linear functionals α_ν on $\mathscr{L}$ and elements e_ν in $\mathscr{L}$ so that for each $g \in \mathscr{L}$ there is some $G \in \mathscr{L}$ with

$$g = G \cdot (f - \lambda) + \sum_{\nu=1}^{k} \alpha_\nu(g) \cdot e_\nu.$$

Fix $q \in \mathcal{M}$ with $f(q) = \lambda$. Then

$$g(q) = \sum_{v=1}^{k} e_v(q)\alpha_v(g),$$

for every $g \in \mathcal{L}$. Hence q, as a functional on $\mathcal{L}$, is a linear combination of $\alpha_1, \ldots, \alpha_k$. It follows that there exist at most k points $q \in \mathcal{M}$ with $f(q) = \lambda$. (Why?) Assertion 3 thus holds.

Recall that $W_l = \{z \in W | f^{-1}(z)$ has l elements$\}$. Because of Assertion 3, $W = \bigcup_{l=1}^{k} W_l$.

Suppose that for some $l < k$, $m(W_l) > 0$, where m is plane Lebesgue measure. Applying Assertion 3 with k replaced by l, we obtain that for each λ in W, $f^{-1}(\lambda)$ has $\leq l$ elements. This contradicts the fact that W_k is nonempty. Hence $m(W_l) = 0$. It follows that $m(W \setminus W_k) = 0$.

Hence each point of W_k is a point of density of W_k, and so we conclude, using Assertion 1, that W_k is open.

Assertion 4. $W \setminus W_k$ is a discrete subset of W.

For each z in W_k denote by

$$p_1(z), \ldots, p_k(z)$$

the points in $f^{-1}(z)$ ordered in some fashion. Fix $z_0 \in W_k$ and choose g in $\mathfrak{A}$ taking distinct values at the points $p_1(z_0), \ldots, p_k(z_0)$. For $z \in W_k$ put

$$\Delta(z) = \prod_{i<j} (g(p_i(z)) - g(p_j(z)))^2.$$

Because of Assertion 1, Δ is analytic in W_k. Also, $\Delta(z_0) \neq 0$.

Let z_1 be a boundary point of W_k in W. Because of Assertion 3, there are less than k points in $f^{-1}(z_1)$. It follows that as $z \to z_1$, $z \in W_k$, $\Delta(z) \to 0$. (Why?) Defining $\Delta = 0$ in $W \setminus W_k$, Δ then becomes continuous on W and analytic on $\{\lambda | \Delta(\lambda) \neq 0\}$.

By Radó's theorem (Theorem 10.6) this implies that Δ is analytic in W. But $\Delta(z_0) \neq 0$, so Δ vanishes on a discrete set. Assertion 4 thus holds.

Fix $p \in f^{-1}(W)$. We must construct a neighborhood of p which is a finite union of analytic disks through p, in the sense of Definition 11.1. If $p \in f^{-1}(W_k)$, Assertion 1 and (8) tell us that we can find a single such disk serving as a neighborhood of p.

For other p, we make use of some elementary constructions involving Riemann surfaces. Put $\zeta = f(p)$ and choose a disk $D = \{z \| z - \zeta| < r\}$ such that $\Delta \neq 0$ in the punctured disk $D' = D \setminus \{\zeta\}$. This is possible since Δ is analytic in W and not identically 0.

Fix $z \in D'$ and define $p_i(z)$ as above. Let $\sigma_1, \ldots, \sigma_k$ be the elementary symmetric functions of $g(p_1(z)), \ldots, g(p_k(z))$. The σ_j are analytic in D' and are bounded there, and so extend to all of D as analytic functions. We have

$$g^k - \sigma_1(f)g^{k-1} + \cdots + (-1)^k\sigma_k(f) = 0 \tag{9}$$

on $f^{-1}(D')$. Consider the equation

$$w^k - \sigma_1(z)w^{k-1} + \cdots + (-1)^k\sigma_k(z) = 0, \tag{10}$$

for $z \in D$. Equation (10) defines a Riemann surface Σ lying over D. We may regard z and w as analytic functions on Σ. Then z provides a k-to-1 covering map of $\Sigma \cap z^{-1}(D')$ on D'.

Fix $\xi \in \Sigma \cap z^{-1}(D')$. $\exists k$ points in $f^{-1}(D')$ where f takes the value $z(\xi)$. At exactly one of these points, say y, $g(y) = w(\xi)$, because of (9) and the fact that $\Delta \neq 0$ in D'. Thus, given ξ, $\exists$ unique $y \in f^{-1}(D')$ with $f(y) = z(\xi)$ and $g(p) = w(\xi)$.

Denote by τ the map: $\xi \to y$ of $\Sigma \cap z^{-1}(D')$ into $f^{-1}(D')$. One easily verifies that τ is continuous, one-to-one, and onto.

Further, $\exists$ subsets $\Lambda_1, \ldots, \Lambda_s$ of Σ with the following properties: The Λ_j are disjoint and $\Sigma = \bigcup_{j=1}^{s} \Lambda_j$. $\exists$ unique point ξ_j in Λ_j with $z(\xi_j) = \zeta$ and z maps $\Lambda_j \cap z^{-1}(D')$ in a finite-sheeted way onto D'. Each Λ_j is conformally equivalent to a disk. Also, as ξ approaches ξ_j, $\tau(\xi)$ approaches a point q in $f^{-1}(\zeta)$, where q depends only on j.

Defining $\tau(\xi_j) = q$ we now extend τ to a continuous one-to-one map of Λ_j into $f^{-1}(D)$. Then $f^{-1}(D) = \bigcup_{j=1}^{s} \tau(\Lambda_j)$.

It is easy to verify that, for $h \in \mathfrak{A}$, $h(\tau)$ is analytic on $\Sigma \cap z^{-1}(D')$. Since h is bounded, it follows that $h(\tau)$ is analytic on all of Σ. Hence each $\tau(\Lambda_j)$ is an analytic disk through $\tau(\xi_j)$.

For certain j, $\tau(\xi_j) = p$. The union of all these $\tau(\Lambda_j)$ is then the desired neighborhood of p.

Theorem 11.2 is thus proved.

NOTES

Theorem 11.2 is essentially contained in the paper by E. Bishop, Holomorphic completions, analytic continuations and the interpolation of semi-norms, *Ann. Math.* **78**, No. 3 (1963). It is not stated explicitly there but is extracted from Section 5 of the paper. The use of Exercise 11.2 in a similar situation is due to T. W. Gamelin.

12

ALGEBRAS OF ANALYTIC FUNCTIONS

Let Ω be a domain in $\mathbf{C}$ and let $\mathfrak{A}_0$ be an algebra of analytic functions on Ω. Assume that $\mathfrak{A}_0$ separates points on Ω and contains the constants.

Let K be a compact set $\subset \Omega$ and write $\mathfrak{A}$ for the closure of $\mathfrak{A}_0$ in the norm

$$\|f\| = \max_K |f|.$$

$\mathfrak{A}$ is thus a uniform algebra on K. Our problem is to describe the maximal ideal space $\mathcal{M}$ of $\mathfrak{A}$.

It can of course occur that $\mathfrak{A} = C(K)$, in which case $\mathcal{M} = K$. This happens, for instance, if Ω is the annulus $a_1 < |z| < a_2$ and K is the circle $|z| = c$, where $a_1 < c < a_2$, and $\mathfrak{A}_0$ consists of all analytic functions on Ω. (Why?)

On the other hand, take Ω again to be the annulus $a_1 < |z| < a_2$ and K to be the circle $|z| = c$, but now take $\mathfrak{A}_0$ to consist of all polynomials in z. Now $\mathfrak{A}$ turns out to consist of all those functions in $C(K)$ that extend analytically to the disk $|z| < c$. Thus $\mathcal{M}$ here is the closed disk $|z| \leq c$.

We shall now study the general case. For simplicity we shall assume K to be the union of a finite number of simple closed analytic curves, or analytic arcs, contained in Ω.

THEOREM 12.1

Let $\mathfrak{A}_0$ be an algebra of analytic functions on Ω and let K be as described. Let $\mathfrak{A}$ denote the uniform closure of $\mathfrak{A}_0$ on K and let $\mathcal{M}$ be the maximal ideal space of $\mathfrak{A}$.

Then each point p in $\mathcal{M} \setminus K$ has a neighborhood in $\mathcal{M}$ which is a finite union of analytic disks through p.

We shall see in Section 13 that this result admits interesting applications to questions of uniform approximation on compact sets in $\mathbf{C}^n$.

We now make the following simplifying assumption, to be removed at the end of the proof:

(I) $\exists$ elements $g_1, \ldots, g_s$ in $\mathfrak{A}_0$ which generate $\mathfrak{A}$.

We shall deduce Theorem 12.1 from the following theorem: For $f \in \mathfrak{A}$ and $z \in \mathbf{C}$ put

$$f^{-1}(z) = \{p \in \mathscr{M} | f(p) = z\}.$$

THEOREM 12.2

Fix $f \in \mathfrak{A}_0$. Then the set of all $z \in \mathbf{C}$ such that $f^{-1}(z)$ is infinite has plane measure 0.

We shall need

Exercise 12.1. Let Ω be a plane domain and fix $z_0 \in \Omega$. For each compact subset K of Ω there exists a constant r, $0 < r < 1$, so that the following holds: If $f \in H(\Omega)$ and $|f| < 1$ on Ω and if f vanishes at z_0 to order λ, then

$$|f| \leq r^\lambda \text{ on } K.$$

Definition. Let f be a polynomial

$$\sum_I c_I z^I$$

in n complex variables $z_1, \ldots, z_n$, where I denotes an n-tuple $(\nu_1, \nu_2, \ldots, \nu_n)$ of nonnegative integers and

$$z^I = z_1^{\nu_1} \cdot z_2^{\nu_2} \cdots z_n^{\nu_n}.$$

We call f a *unit polynomial* if $\max_I |c_I| = 1$.

We say f is *of degree* $(d_1, \ldots, d_n)$ if for all $I = (\nu_1, \ldots, \nu_n)$, $\nu_j \leq d_j$, all j.

LEMMA 12.3

Let f be a unit polynomial in one complex variable z of degree k and let α be a positive number. Put

$$Q = \{z \,|\, |z| \leq 1, |f(z)| \leq \alpha^k\}.$$

Then $m(Q) \leq 48\alpha$, where m denotes plane Lebesgue measure.

Proof. Let $\zeta_1, \ldots, \zeta_t$ be the roots of f with modulus ≤ 2 and $\zeta_{t+1}, \ldots, \zeta_k$ the remaining roots. Then without loss of generality

(1) $$f(z) = C(z - \zeta_1) \cdots (z - \zeta_t)(1 - (\zeta_{t+1})^{-1}z) \cdots (1 - (\zeta_k)^{-1}z).$$

Hence for $|z| = 1$,

$$|f(z)| \leq |C|(1 + |\zeta_1|) \cdots (1 + |\zeta_t|)(1 + |\zeta_{t+1}|^{-1}) \cdots (1 + |\zeta_k|^{-1}) \leq |C| 3^k.$$

It follows that the modulus of each coefficient of f is $\leq |C|3^k$. (Why?)

Since f is a unit polynomial we conclude that

$$1 \le |C|3^k \quad \text{or} \quad |C|^{-1} \le 3^k. \tag{2}$$

Also, by (1) we have for $z \in Q$,

$$|C||z - \zeta_1| \cdots |z - \zeta_t| \le \alpha^k |1 - (\zeta_{t+1})^{-1} z|^{-1} \cdots |1 - (\zeta_k)^{-1} z|^{-1}$$
$$\le \alpha^k \cdot 2^{k-t} \le 2^k \alpha^k.$$

Combining this with (2) we get

$$|z - \zeta_1| \cdots |z - \zeta_t| \le (6\alpha)^k \qquad \text{for } z \in Q.$$

Put Q_0 = projection of Q on the x-axis. Put $\alpha_j = \operatorname{Re} \zeta_j$, $1 \le j \le t$. Then for $x \in Q_0$,

$$|x - \alpha_1| \cdots |x - \alpha_t| \le (6\alpha)^k. \tag{3}$$

We now appeal to the following:

***Exercise 12.2.** Let $\alpha_1, \ldots, \alpha_t$ be real numbers and $P(x) = (x - \alpha_1) \cdots (x - \alpha_t)$. Fix M and put

$$S = \{x \,|\, |P(x)| \le M\}.$$

Then the linear measure of $S \le 4 \cdot M^{1/t}$.

Let μ denote linear measure. The exercise applied to (3) gives

$$\mu(Q_0) \le (6\alpha)^{k/t} \cdot 4.$$

For $\alpha < \frac{1}{6}$, this gives $\mu(Q_0) \le 24\alpha$. For $\alpha \ge \frac{1}{6}$, $\mu(Q_0) \le 2 \le 24\alpha$. So $\mu(Q_0) \le 24\alpha$, and we conclude that $m(Q) \le 48\alpha$. Q.E.D.

LEMMA 12.4

Let Ω be a plane domain and K a compact subset. Let $\mathfrak{A}_0$ be an algebra of analytic functions on Ω. Put $\|\phi\| = \max_K |\phi|$, all $\phi \in \mathfrak{A}_0$.

Fix $f, g \in \mathfrak{A}_0$. Then there exists r, $0 < r < 1$, and $c > 0$ such that for each pair of positive integers (d, e) we can find a unit polynomial $F_{d,e}$ in two variables of degree (d, e) such that

$$\|F_{d,e}(f, g)\| \le c^{d+e} \cdot r^{de}.$$

Proof. Choose a subregion Ω_1 of Ω, with $K \subset \Omega_1$ and $\bar{\Omega}_1$ a compact subset of Ω. Choose $c_0 > 1$ with $|f| < c_0$, $|g| < c_0$ on Ω_1. Also fix some $z_0 \in \Omega_1$. Consider an arbitrary polynomial

$$F(z, w) = \sum_{n=0}^{d} \sum_{m=0}^{e} c_{nm} z^n w^m.$$

Put $h = F(f, g)$.

The requirement that h should vanish at z_0 of order λ imposes λ linear homogeneous conditions on the c_{nm} and hence has a nontrivial solution if $\lambda < (d + 1) \cdot (e + 1)$.

We may assume that the corresponding polynomial $F = F_{d,e}$ is a unit polynomial. Now

$$\frac{d^\nu h}{dz^\nu}(z_0) = 0, \qquad \nu = 0, 1, 2, \ldots, \lambda - 1.$$

Exercise 12.1 applied to h, Ω_1, and z_0 gives

$$|h| \leq \max_{\Omega_1} |h| \cdot r^\lambda \text{ on } K.$$

On Ω_1,

$$|h| \leq \sum_{n,m} |c_{nm}||f|^n|g|^m \leq (d + 1)(e + 1) \cdot c_0^{d+e}.$$

Hence, for large c,

$$\|h\| \leq (d + 1)(e + 1)c_0^{d+e}r^\lambda \leq c^{d+e}r^\lambda \qquad \text{for all } d, e. \qquad \text{Q.E.D.}$$

Proof of Theorem 12.2. Fix $g \in \mathfrak{A}_0$ and put, for each $z \in \mathbf{C}$, $S(z) = \{w \in \mathbf{C} | \exists m \in \mathcal{M}$ with $m(f) = z$ and $m(g) = w\}$. Thus $S(z)$ is the set of values g takes on $f^{-1}(z)$.

By Lemma 12.4, for each (d, e) there is a unit polynomial F of degree (d, e) and $r, 0 < r < 1$, and c, independent of d, e, so that

$$\|F(f, g)\| \leq c^{d+e} \cdot r^{de}. \tag{4}$$

Fix r_0 with $r < r_0 < 1$. Then there exists d_0 with

$$c^{d+e} \cdot r^{de} \leq r_0^{de} \qquad \text{for } d,e > d_0.$$

Henceforth we assume that $d,e > d_0$.

Fix $z \in \mathbf{C}$ and $w \in S(z)$. Equation (4) gives

$$|F(z, w)| \leq r_0^{de}. \tag{5}$$

(Why?)

Now

$$F(z, w) = \sum_{j=0}^{e} G_j(z)w^j.$$

Without loss of generality we can suppose that $\|f\| \leq 1$, so that $S(z)$ is empty unless $z \in D = \{z | |z| \leq 1\}$.

Since F is a unit polynomial, $\exists j$ such that $G = G_j$ is a unit polynomial of degree d. Put

$$T(d, e) = \{z \in D | |G(z)| \leq r_0^{ed/2}\}.$$

By Lemma 12.3 with $\alpha = r_0^{e/2}$, $m(T(d, e)) \leq 48r_0^{e/2}$.

Assertion 1. If $z_1 \in D \setminus T(d, e)$, then $\exists$ unit polynomial B of degree e with

$$|B(w)| \leq r_0^{de/2} \qquad \text{for all } w \in S(z_1).$$

For put

$$A(w) = F(z_1, w),$$

where F is as above. Then A is a polynomial of degree e having one coefficient, $G(z_1)$, with

$$(6) \qquad |G(z_1)| > r_0^{ed/2}.$$

Also, by (5)

$$(7) \qquad |A(w)| \leq r_0^{de} \qquad \text{if } w \in S(z_1).$$

Dividing A by the modulus of its largest coefficient, we get a polynomial B with the properties claimed in Assertion 1. For each e put

$$H_e = \{z | z \in D \setminus T(d, e) \text{ for infinitely many } d\}.$$

Assertion 2. Fix e and fix $z^* \in D$. If $z^* \in H_e$, then $S(z^*)$ has at most e elements.

Since $z^* \in H_e$, for arbitrarily large d, $z^* \in D \setminus T(d, e)$. Let $\{d_j\}$ be such a sequence of d's, $\to \infty$.

By Assertion 1, for each $j \exists$ unit polynomial B_j of degree e with

$$(8) \qquad |B_j(w)| \leq r_0^{d_j e/2} \qquad \text{for each } w \in S(z^*).$$

Since the B_j have fixed degree e and are unit polynomials, there is a subsequence of $\{B_j\}$ converging uniformly on D to a polynomial B^*. Then B^* is a unit polynomial of degree e. Since $d_j \to \infty$ and $r_0 < 1$, (8) yields

$$B^*(w) = 0 \qquad \text{for each } w \in S(z^*).$$

It follows that $S(z^*)$ contains at most e elements, proving Assertion 2.

It follows that for every e,

$$\{z \in \mathbf{C} | S(z) \text{ is infinite}\} \subset D \setminus H_e.$$

Assertion 3. $m(D \setminus H_e) \leq 48 r_0^{e/2}$.

To prove Assertion 3, observe that if $z \in D \setminus H_e$, then $z \in T(d, e)$, for all d from some point on, so

$$(9) \qquad D \setminus H_e \subseteq \bigcup_{k=d_0}^{\infty} \bigcap_{d=k}^{\infty} T(d, e).$$

But for each fixed d,

$$m(T(d, e)) \leq 48 r_0^{e/2}.$$

Since the right-hand side in (9) is the union of an increasing family of sets, each set being of measure $\leq 48 r_0^{e/2}$, we get $m(D \setminus H_e) \leq 48 r_0^{e/2}$, proving the assertion.

Put $S_{f,g} = \{z | S(z) \text{ is infinite}\}$. Then $S_{f,g} \subset D \setminus H_e$ for each e, so by Assertion 3, $m(S_{f,g}) \leq 48 r_0^{e/2}$. Since $r_0 < 1$ and e is arbitrary, $m(S_{f,g}) = 0$.

This now holds for each $g \in \mathfrak{A}_0$, in particular for the generators $g_1, \ldots, g_s$ of (I).

Fix z_0 with $f^{-1}(z_0)$ infinite. Since $g_1, \ldots, g_s$ together separate points, $\exists j$ such that g_j takes infinitely many values on $f^{-1}(z_0)$; i.e., $z_0 \in S_{f,g_j}$. Thus

$$\{z | f^{-1}(z) \text{ is infinite}\} \subset \bigcup_{j=1}^{s} S_{f,g_j}.$$

Since each summand on the right has measure 0, we conclude that

$$m\{z|f^{-1}(z) \text{ is infinite}\} = 0. \qquad \text{Q.E.D.}$$

Proof of Theorem 12.1. Fix $p_0 \in \mathcal{M} \setminus K$. We shall exhibit a neighborhood of p_0 in $\mathcal{M}$ which is a finite union of analytic disks through p_0.

Assertion. There is an $f \in \mathfrak{A}_0$ so that $f(p_0) \in \mathbf{C} \setminus f(K)$.

Choose $f_1 \in \mathfrak{A}_0, f_1$ not a constant. Then

$$\{p \in K | f_1(p) = f_1(p_0)\}$$

is a finite set. Choose $f_2 \in \mathfrak{A}_0$ with $f_2 \neq f_2(p_0)$ on this set. Define

$$Q(z) = \frac{f_1(z) - f_1(p_0)}{f_2(z) - f_2(p_0)}.$$

Restricted to K, Q, being meromorphic, omits some value η.

Put $f = f_1 - \eta f_2 \in \mathfrak{A}_0$. If $f(K) \ni f(p_0)$, then for some $z \in K$,

$$f_1(z) - \eta f_2(z) = f_1(p_0) - \eta f_2(p_0),$$

so

$$f_1(z) - f_1(p_0) = \eta(f_2(z) - f_2(p_0)).$$

If $f_2(z) - f_2(p_0) = 0$, then also $f_1(z) = f_1(p_0)$, contrary to the choice of f_2. So $Q(z) = \eta$, contrary to the choice of η. Hence f satisfies the assertion.

Now let W be the component of $\mathbf{C} \setminus f(K)$ containing $f(p_0)$. By Theorem 12.2, $f^{-1}(z)$ is a finite set for almost all z in W. Theorem 11.2 now yields that p_0 has a neighborhood as desired. This proves Theorem 12.1, under hypothesis (I). Now drop (I).

Fix $f \in \mathfrak{A}_0$ and pick $z \in \mathbf{C} \setminus f(K)$. Suppose that $f^{-1}(z)$ is infinite; i.e., $\exists \{p_j\}_{j=1}^{\infty}$ in $\mathcal{M}$ with $f(p_j) = z$, all j. Without loss of generality, $\mathfrak{A}_0$ is closed under uniform convergence on some compact set containing K in its interior. Then $\exists g \in \mathfrak{A}_0$ such that g separates points on $\{p_j\}$. (Why?)

Let $\mathfrak{A}_1$ be the algebra of all polynomials in f and g, and $\overline{\mathfrak{A}_1}$ its uniform closure on K. Then $\mathfrak{A}_1$ satisfies (I). Hence the assertion of Theorem 12.1 is valid for $\mathfrak{A}_1$. This implies in particular that

$$\Lambda = \{p \in \mathcal{M}(\overline{\mathfrak{A}_1}) | f(p) = z\}$$

is a finite set. But each p_j induces a point of $\mathcal{M}(\overline{\mathfrak{A}_1})$, and since g separates the p_j, these points are distinct. Thus Λ is infinite, which is a contradiction. Hence $f^{-1}(z)$ was finite, and so Theorem 11.2 applies as before. Thus hypothesis (I) was irrelevant.

Q.E.D.

NOTES

Theorem 12.1 in the special case when Ω is an annulus and K a concentric circle in Ω is due the author, The maximum principle for bounded functions, *Ann. Math.* **69**, No. 3 (1959). In a more general form than the one given here, it is proved in E. Bishop,

Analyticity in certain Banach algebras, *Trans. Am. Math. Soc.* **102** (1962), and in Holomorphic completions, analytic continuations and the interpolation of semi-norms, *Ann. Math.* **78** (1963). The proof given here is the one in the latter paper. An independent proof of the theorem is due to H. Royden, Algebras of bounded analytic functions on Riemann surfaces, *Acta Math.* **114** (1965).

13

APPROXIMATION ON CURVES IN $\mathbf{C}^n$

Let K be a simple closed curve (homeomorphic image of a circle) in $\mathbf{C}^n$, and J a Jordan arc (homeomorphic image of a closed interval) in $\mathbf{C}^n$. We wish to describe the polynomially convex hulls $h(K)$ and $h(J)$.

For $n = 1$ this description can be given at once. (What is it?)

For $n > 1$ we need the notion of an "analytic variety," which, roughly speaking, is a subset of $\mathbf{C}^n$ locally definable by analytic equations.

Definition 13.1. Let Ω be an open set in $\mathbf{C}^n$ and V a relatively closed subset of Ω. V is an *analytic subvariety* of Ω if for each $x^0 \in V$ we can find a neighborhood N of x^0 in $\mathbf{C}^n$ and functions $\phi_1, \ldots, \phi_s \in H(N)$ such that

$$V \cap N = \{x \in N | \phi_j(x) = 0, j = 1, \ldots, s\}.$$

The reader may consult the book of Gunning and Rossi [6, Chaps. II and III], for a good introduction to the subject of analytic varieties. Here we shall only encounter analytic varieties whose complex dimension equals 1.

Definition 13.2. Let V be as above and fix $x^0 \in V$. Assume that after some analytic change of coordinates in a neighborhood N of x^0 to new coordinates $Z_1, \ldots, Z_n$, we have for some integer k,

$$V \cap N = \{x \in N | Z_j(x) = 0, j = 1, 2, \ldots, n - k\}.$$

Then x^0 is a *regular* point on V. k is independent of the choice of coordinates and is the *dimension of* V *at* x^0. We say that V has *dimension* k if at each regular point of V the dimension is k.

Note that the set of regular points on a one-dimensional analytic subvariety V constitutes a (possibly disconnected) Riemann surface, and that the complement of this set in V is a discrete subset of V.

Now choose a and b with $a < 1 < b$ and let W be the annulus $a < |z| < b$. Let Γ be the circle $|z| = 1$. Let $f_1, \ldots, f_n$ be functions analytic in W and assume the f_j

together separate points on W. Denote by K the image in $\mathbf{C}^n$ of Γ under the map $z \to (f_1(z), \ldots, f_n(z))$.

A set K obtained in this way we call an *analytic curve* in $\mathbf{C}^n$.

THEOREM 13.1

Let K be an analytic curve in $\mathbf{C}^n$. Then $h(K) \setminus K$ is a one-dimensional analytic subvariety of some open set in $\mathbf{C}^n$.

Note. $h(K) \setminus K$ may be empty. (Give an example.)

Proof. Let $\mathfrak{A}_0$ be the algebra of all polynomials in $f_1, \ldots, f_n$, regarded as an algebra of functions on W. Let $\mathfrak{A}$ denote the uniform closure of $\mathfrak{A}_0$ on Γ. Theorem 12.1 then applies and yields that if $\mathcal{M} = \mathcal{M}(\mathfrak{A})$, then every point in $\mathcal{M} \setminus \Gamma$ has a neighborhood in $\mathcal{M}$ which is a finite union of analytic disks. (Recall Definition 11.1.)

Each f_j may be regarded as a continuous function on $\mathcal{M}$, and the f_j together separate points on $\mathcal{M}$.

The image of $\mathcal{M}$ in $\mathbf{C}^n$ under the map $f: M \to (f_1(M), \ldots, f_n(M))$ is $h(K)$. (Why?)

Fix $x^0 \in h(K) \setminus K$. Then $\exists M^0 \in \mathcal{M} \setminus \Gamma$ with $f(M^0) = x^0$. Let U be a neighborhood of M^0 in $\mathcal{M}$ which is a finite union of analytic disks, $D_1, \ldots, D_k$. Then the union of the $f(D_j)$, $j = 1, \ldots, k$, is a neighborhood U^* of x^0 in $h(K)$.

Fix j, $1 \leq j \leq k$. By the definition of an analytic disk, $\exists$ a one-to-one continuous map Φ from $|\lambda| < 1$ onto D_j such that $h(\Phi)$ is analytic in $|\lambda| < 1$ for every $h \in \mathfrak{A}$. Define $z_i(\lambda) = f_i(\Phi(\lambda))$, $i = 1, \ldots, n$. Then the map $\lambda \to (z_1(\lambda), \ldots, z_n(\lambda))$ is a one-to-one continuous map of $|\lambda| < 1$ onto $f(D_j)$, and each z_i is an analytic function in $|\lambda| < 1$.

Denote by $B(x, r)$ the open ball in $\mathbf{C}^n$ of center x and radius r.

***Exercise 13.1.** Fix j and put $D = D_j$. For small $\varepsilon > 0$, $f(D) \cap B(x^0, \varepsilon)$ is a one-dimensional analytic subvariety of $B(x^0, \varepsilon)$.

Exercise 13.2. For small $\varepsilon > 0$, $U^* \cap B(x^0, \varepsilon)$ is a one-dimensional analytic subvariety of $B(x^0, \varepsilon)$.

Let Ω be the union of all the balls $B(x^0, \varepsilon)$ obtained in Exercise 13.2 as x^0 varies over $h(K) \setminus (K)$. We may assume that each $B(x^0, \varepsilon)$ is disjoint from K. Hence $h(K) \setminus K$ is a closed subset of Ω, and, by choice of the $B(x^0, \varepsilon)$, $h(K) \setminus K$ is a one-dimensional analytic subvariety of Ω. Q.E.D.

Now let W be an open neighborhood in $\mathbf{C}$ of the unit interval $I: 0 \leq x \leq 1$, let $f_1, \ldots, f_n$ be functions analytic in W, and assume that the f_j together separate points on W. Denote by J the image in $\mathbf{C}^n$ of I under the map $z \to (f_1(z), \ldots, f_n(z))$.

A set J obtained in this way we call an *analytic arc* in $\mathbf{C}^n$.

THEOREM 13.2

Let J be an analytic arc in $\mathbf{C}^n$. Then J is polynomially convex.

COROLLARY

Let J be an analytic arc in $\mathbf{C}^n$. Then $P(J) = C(J)$.

Definition 13.3. Let X be a compact set in $\mathbf{C}^n$. $R_0(X)$ is the set of all $f \in C(X)$ such that $\exists$ polynomials A and B, with $B \neq 0$ on X and $f = A/B$ on X.

LEMMA 13.3

Let X be a compact set in $\mathbf{C}^n$ such that $R_0(X)$ is dense in $C(X)$. Fix $x^0 \in \mathbf{C}^n \setminus X$. Then $\exists$ polynomial F with $F(x^0) = 0$ and $F \neq 0$ on X.

Proof. Suppose not. Then every polynomial vanishing at x^0 has a zero on X. So if $f = A/B \in R_0(X)$, then $B(x^0) \neq 0$. We define a map $\phi : f \to A(x^0)/B(x^0)$ from $R_0(X) \to \mathbf{C}$. ϕ is well defined and $|\phi(f)| \leq \max_X |f|$. (Why?) ϕ is then a bounded homomorphism of $R_0(X) \to \mathbf{C}$ and so extends to a homomorphism of $C(X) \to \mathbf{C}$. Hence $\exists x^1 \in X$ with $\phi(f) = f(x^1)$, all $f \in C(X)$. Letting f be the coordinate functions z_j, we conclude that $x^0 = x^1$. This is a contradiction. Q.E.D.

LEMMA 13.4

Let J be an analytic arc in $\mathbf{C}^n$. Then $R_0(J)$ is dense in $C(J)$.

Proof. Fix a coordinate function z_k and put $S = z_k(J)$. Since S is the analytic image of an interval, S has plane measure 0. It follows by Theorem 2.8 that $\exists$ rational functions f_ν with poles off S and $\lim_{\nu\to\infty} f_\nu(\zeta) = \bar{\zeta}$ uniformly on S.

For each ν, $f_\nu \circ z_k \in R_0(J)$. Hence $\bar{z}_k$ lies in the uniform closure of $R_0(J)$ on J. By the Stone–Weierstrass theorem we get the assertion.

Proof of Theorem 13.2. Put $V = h(J) \setminus J$. We must show that $V = \varnothing$.

The proof of Theorem 13.1 applies to analytic arcs just as well as to analytic curves, and so V is an analytic subvariety of an open set Ω in $\mathbf{C}^n$.

Fix a regular point $x^0 \in V$. By Lemmas 13.3 and 13.4, $\exists$ polynomial F with $F \neq 0$ on J and $F(x^0) = 0$.

Since J is a smooth arc, J has arbitrarily small simply connected neighborhoods in $\mathbf{C}^n$. (Why?) Choose such a neighborhood U with $F \neq 0$ in U. Then $\exists H \in C(U)$ with $F = e^H$ in U.

Next define V_ε as the subset of V consisting of all points whose distance from $J \geq \varepsilon$. V_ε is compact and its boundary in V lies in U, if ε is small. Since V is a one-dimensional analytic variety, it is easily seen that we can find a compact set W_ε with $V_\varepsilon \subset W_\varepsilon \subset V$ such that the boundary of W_ε in V, ∂W_ε, is a finite union of simple closed curves and $\partial W_\varepsilon \subset U$. For small ε, $x^0 \in W_\varepsilon$.

Exercise 13.3. Let V be a one-dimensional analytic subvariety of a region Ω in $\mathbf{C}^n$ and $\phi \in H(\Omega)$. Let W be an open subset of V whose boundary ∂W is a finite union of simple closed curves oriented positively with respect to W.

Show the following: If $x^0 \in W$ and x^0 is a regular point of V and if $\phi(x^0) = 0$ and $\phi \neq 0$ on ∂W, then $\text{var}\arg_{\partial W}\phi \neq 0$.

Hint. Make a suitable triangulation of W and use the argument principle on each triangle.

Recall that $F = e^H$ in U and hence on ∂W_ε. It follows that $\text{var}\arg_{\partial W_\varepsilon} F = 0$, whence, by Exercise 13.3, F cannot vanish at x^0. This contradiction shows that $V = \varnothing$, as desired.

Proof of Corollary. Since J is polynomially convex and every $f \in R_0(J)$ is holomorphic in a neighborhood of J, $R_0(J) \subset P(J)$ by the Oka–Weil theorem. But $R_0(J)$ is dense in $C(J)$ by Lemma 13.4. Hence $P(J) = C(J)$, as claimed.

For an arc J in the complex plane, $P(J) = C(J)$ whether or not J is analytic, by Theorem 2.11. For $n > 1$, Theorem 13.2 no longer holds for arbitrary arcs (homeomorphic images of a closed interval) in $\mathbf{C}^n$. We shall give an example in $\mathbf{C}^3$.

If γ is an arc in the plane, denote by $\mathfrak{A}_\gamma$ the algebra of functions continuous on the Riemann sphere S^2 and analytic on $S^2 \setminus \gamma$. For smooth arcs γ, $\mathfrak{A}_\gamma$ reduces to the constants. It is not clear, a priori, whether or not there exist arcs γ such that $\mathfrak{A}_\gamma$ contains nonconstant functions.

LEMMA 13.5

If γ has positive plane measure, then $\mathfrak{A}_\gamma$ contains enough functions to separate points on S^2; in fact, three functions in $\mathfrak{A}_\gamma$ do so.

Note. To obtain an arc γ having positive plane measure one can proceed like this: Choose a compact, totally disconnected set E on the real line, having positive linear measure. $E \times E$ is then a compact, totally disconnected subset of $\mathbf{R}^2$ having positive plane measure. Through every compact, totally disconnected subset of the plane an arc may be passed, as was shown by F. Riesz [Sur les ensembles discontinus, *Compt. Rend.* **141** (1905), 650]; γ can be such an arc.

The first example of an arc of positive plane measure was found by Osgood in 1903 by a different method.

Proof. Put

$$F(\zeta) = \int_\gamma \frac{dx\,dy}{z - \zeta}.$$

$F(\zeta) \to 0$ as $\zeta \to \infty$ and $\lim_{\zeta \to \infty} \zeta \cdot F(\zeta) \neq 0$. (Why?) Hence F is not a constant. Fix $\zeta_0 \in \gamma$. The integral defining $F(\zeta_0)$ converges absolutely. (Why?) We claim that F is continuous at ζ_0. For put

$$g(z) = \begin{cases} \dfrac{1}{z}, & |z| < R, \\ 0, & |z| > R \end{cases}$$

for R some large number. Then $g \in L^1(\mathbf{R}^2)$.

$$|F(\zeta) - F(\zeta_0)| \le \int_\gamma \left| \frac{1}{z - \zeta} - \frac{1}{z - \zeta_0} \right| dx\,dy$$

$$= \int_\gamma |g(z - \zeta) - g(z - \zeta_0)|\, dx\,dy \to 0 \quad \text{as} \quad \zeta \to \zeta_0,$$

since $g \in L^1(\mathbf{R}^2)$. Hence the claim is established. Thus $F \in C(S^2)$, and since F evidently is analytic on $S^2 \setminus \gamma$, $F \in A_\gamma$.

Fix now $a,b \in S^2 \setminus \gamma$ with $F(a) \neq F(b)$. Then $F_2, F_3 \in A_\gamma$, where

$$F_2(z) = \frac{F(z) - F(a)}{z - a}, \qquad F_3(z) = \frac{F(z) - F(b)}{z - b}.$$

Fix distinct points $z_1, z_2 \in S^2$. It is easily checked that if $F(z_1) = F(z_2)$, then either F_2 or F_3 separates z_1 and z_2. Hence F, F_2, and F_3 together separate points on S^2.

Q.E.D.

We now define an arc J_0 in $\mathbf{C}^3$ as the image of a given plane arc γ having positive plane measure under the map $\zeta \to (F(\zeta), F_2(\zeta), F_3(\zeta))$.

THEOREM 13.6

J_0 is not polynomially convex in $\mathbf{C}^3$. Hence $P(J_0) \neq C(J_0)$.

Proof. Fix $\zeta_0 \in S^2 \setminus \gamma$. Then $x^0 = (F(\zeta_0), F_2(\zeta_0), F_3(\zeta_0)) \notin J_0$. Yet if P is any polynomial on $\mathbf{C}^3$,

$$|P(x^0)| \leq \max_{J_0} |P|.$$

For $f = P(F, F_2, F_3) \in A_\gamma$, so by the maximum principle

$$|f(\zeta_0)| \leq \max_{\gamma} |f|,$$

as asserted. Hence $x^0 \in h(J_0) \setminus J_0$, and we are done.

***Exercise 13.4.** If ϕ is a nonconstant element of $P(J_0)$, then $\phi(J_0)$ is a Peano curve in $\mathbf{C}$, i.e., contains interior points. In particular, the coordinate projections of J_0, $z_k(J_0)$, are points or Peano curves.

NOTES

Theorems 13.1 and 13.2 were first proved by the author, The hull of a curve in $\mathbf{C}^n$, *Ann. Math.* **68** (1958). They were generalized and given new proofs by E. Bishop and H. Royden in the papers cited in the Notes for Section 12. The hypothesis of analyticity was weakened to differentiability by E. Bishop (unpublished). A proof for the differentiable case was given by G. Stolzenberg, Uniform approximation on smooth curves, *Acta Math.* **115** (1966). We have followed Stolzenberg in the proof of Theorem 13.2. Various other generalizations of Theorems 13.1 and 13.2 have also more recently been given by other authors, in particular by L. A. Markusevic, J.-E. Björk, and H. Alexander. Lemma 13.4, for differentiable curves, is due to H. Helson and F. Quigley, Existence of maximal ideals in algebras of continuous functions, *Proc. Am. Math. Soc.* **8** (1957).

The example of an arc J_0 in $\mathbf{C}^3$ which is not polynomially convex is due to the author, Polynomial approximation on an arc in $\mathbf{C}^3$, *Ann. Math.* **62** (1955). The nonconstant function F in A_γ used in Lemma 13.5 was found by Denjoy. A modification of the construction of J_0 which provides a nonpolynomially convex arc in $\mathbf{C}^2$ is due to W. Rudin, Subalgebras of spaces of continuous functions, *Proc. Am. Math. Soc.* **7** (1956). The fact that for every arc $J \subset \mathbf{C}$, $P(J) = C(J)$, was first proved by J. L. Walsh in 1926.

14

UNIFORM APPROXIMATION ON DISKS IN $\mathbf{C}^n$

As the two-dimensional analogue of an arc in $\mathbf{C}^n$, we take a disk in $\mathbf{C}^n$ defined as follows. Let D be the closed unit disk in the ζ-plane and let $f_1, \ldots, f_n$ be continuous functions defined on D. Assume that the map $\zeta \to (f_1(\zeta), \ldots, f_n(\zeta))$ is one to one on D. The image $\tilde{D}$ of D under this map we call a *disk* in $\mathbf{C}^n$.

Our problem is to give conditions on $\tilde{D}$ in order that $P(\tilde{D}) = C(\tilde{D})$. A necessary condition is

(1) $$\tilde{D} \text{ is polynomially convex in } \mathbf{C}^n.$$

Condition (1) is clearly not sufficient.

We make the simplifying assumption that

(2) $$f_1(\zeta) = \zeta.$$

When $f_2, \ldots, f_n$ are merely assumed continuous, we do not know sufficient conditions. Let us now suppose that the f_j have continuous first partials in a neighborhood of D; i.e., $\zeta = \xi + i\eta$ and $\partial f_j/\partial\xi$ and $\partial f_j/\partial\eta$ exist and are continuous. Then

THEOREM 14.1

Assume (1) *and* (2), *and that*

(3) $$\text{for every } \zeta_0 \in D, \frac{\partial f_j}{\partial \bar{\zeta}}(\zeta_0) \neq 0 \text{ for some } j.$$

Then $P(\tilde{D}) = C(\tilde{D})$.

Note. The conclusion is equivalent to saying that the space of polynomials in $f_1, f_2, \ldots, f_n$ is dense in $C(D)$. Observe that if (3) fails for each ζ_0 in some open subset of D, then every function that is a uniform limit on D of polynomials in $f_1, \ldots, f_n$ is analytic there, whence $P(\tilde{D}) \neq C(\tilde{D})$. Condition (3) is thus a natural restriction.

LEMMA 14.2

Fix a in D. $\exists$ a neighborhood Ω of $\tilde{D}$ in $\mathbf{C}^n$ and $\exists h \in H(\Omega)$ such that

(4) $$h = (z_1 - a)h_1 \text{ in } \Omega \text{ with } h_1 \in H(\Omega), \text{ and}$$

(5) $$\exists \text{ a circular sector } T : -\frac{\pi}{4} \leq \theta \leq \frac{\pi}{4}, 0 \leq r \leq \varepsilon \text{ such that } h(\tilde{D}) \cap T = \{0\}.$$

Proof. By (3), $\exists f = f_j$, $1 \leq j \leq n$, with $f_{\bar\zeta}(a) = \partial f/\partial\bar\zeta(a) \neq 0$. If $\zeta \in D$,

$$f(\zeta) = f(a) + f_\zeta(a)(\zeta - a) + f_{\bar\zeta}(a)(\bar\zeta - \bar a) + O(|\zeta - a|).$$

Then

$$\frac{(\zeta - a)(f(\zeta) - f(a) - f_\zeta(a)(\zeta - a))}{f_{\bar\zeta}(a)} = |\zeta - a|^2 + O(|\zeta - a|^2).$$

Put

$$g(z_1, \ldots, z_n) = -(z_1 - a)\left(\frac{z_j - f(a) - f_\zeta(a)(z_1 - a)}{f_{\bar\zeta}(a)}\right).$$

Thus for $\zeta \in D$

$$g(f_1(\zeta), \ldots, f_n(\zeta)) = -|\zeta - a|^2 + O(|\zeta - a|^2).$$

Hence g is a polynomial such that

(6) $$g(y) = 0, \qquad \text{where } y = (f_1(a), f_2(a), \ldots, f_n(a)),$$

and

(7) $\operatorname{Re} g(x) < 0$ for $x \in \tilde{D} \setminus \{y\}$ and x in some neighborhood U of y in $\mathbf{C}^n$.

Because of (7) we can find an open set V in C^n such that

$$U \cup V \supset \tilde{D} \qquad \text{and} \qquad \operatorname{Re} g < 0 \text{ in } U \cap V.$$

Also $\tilde{D}$ is polynomially convex by (1). By Lemma 9.4, $\exists \phi \in H(U)$, $\psi \in H(V)$ with

$$\log g = \psi - \phi \text{ in } U \cap V,$$

where we may have had to shrink U and V a bit first. Hence $ge^\phi = e^\psi$ in $U \cap V$. The right side $\in H(V)$ and the left side $\in H(U)$. Hence the function h defined by

$$h = e^\psi \text{ in } V, \qquad h = ge^\phi \text{ in } U$$

is holomorphic in $U \cup V$. Put $k = e^\phi$. Then

(8) $$h = kg \text{ in } U, k \in H(U) \text{ and } k \neq 0 \text{ there.}$$

(9) $$h \neq 0 \qquad \text{on } \tilde{D} \setminus \{y\}.$$

Since $z_1 - a \neq 0$ on $\tilde{D}$ except at y, we may assume without loss of generality that $z_1 - a \neq 0$ in V. Put

$$h_1 = \frac{h}{z_1 - a}.$$

Then $h_1 \in H(V)$. Also $g/(z_1 - a)$ is a polynomial, and so by (8) $h_1 \in H(U)$. Hence $h_1 \in H(U \cup V)$.

Putting $\Omega = U \cup V$, we see that (4) holds. Without loss of generality, $k(y) = 1$. By (8), $h - g = (k - 1)g$, whence

$$(10) \qquad |h - g| < \frac{1}{\sqrt{2}}|g|$$

in some neighborhood U_1 of y with $U_1 \subset U$, except at y. Fix x in U_1 and $x \neq y$. By (7) $\operatorname{Re} g(x) \leq 0$. Hence for w in the sector $-\pi/4 \leq \arg w \leq \pi/4$, $|w - g(x)| \geq (1/\sqrt{2})|g(x)|$. Because of (10), this means that $h(x)$ lies outside the sector.

On the other hand, $\tilde{D} \setminus U_1$ is a compact subset of $\tilde{D}$ avoiding y. By (9) $h \neq 0$ on $\tilde{D} \setminus U_1$, and so for some $\varepsilon > 0$, $|h| > \varepsilon$ on $\tilde{D} \setminus U_1$. Hence everywhere on $\tilde{D} \setminus \{y\}$ the value of h lies outside the sector T given in (5), whence (5) holds. Q.E.D.

Denote by $\mathfrak{A}$ the uniform closure on D of the algebra of polynomials in $f_1, \ldots, f_n$. Again fix a in D.

LEMMA 14.3

$\exists \phi_n \in \mathfrak{A}$, $n = 1, 2, \ldots$ with

$$(11) \qquad \lim_{n \to \infty} \phi_n(\zeta) = \frac{1}{\zeta - a}, \qquad \zeta \in D \setminus \{a\}.$$

$$(12) \qquad |\phi_n(\zeta)| \leq \frac{b}{|\zeta - a|}, \qquad \textit{all } \zeta \in D, \textit{ all } n, \textit{ where } b \textit{ is a constant.}$$

Proof. With h and h_1 as in (4), put

$$\psi_n(z) = h_1(z) \frac{1}{h(z) - 1/n}.$$

Since h_1 and h are holomorphic in a neighborhood of $\tilde{D}$ and $h(z) \neq 1/n$ in some neighborhood of $\tilde{D}$ for n large, by (5), ψ_n is holomorphic in a neighborhood of $\tilde{D}$ for n large. Hence the restriction of ψ_n to $\tilde{D}$ lies in $P(\tilde{D})$, using the fact that $\tilde{D}$ is polynomially convex. Put $f(\zeta) = (f_1(\zeta), \ldots, f_n(\zeta))$. It follows that $\psi_n(f) \in \mathfrak{A}$. Put $\phi_n(\zeta) = \psi_n(f(\zeta))$, $\zeta \in D$. Thus $\phi_n \in \mathfrak{A}$. Also for $\zeta \in D$, $\zeta \neq a$,

$$\lim_{n \to \infty} \phi_n(\zeta) = \lim_{n \to \infty} h_1(f(\zeta)) \frac{1}{(\zeta - a)h_1(f(\zeta)) - 1/n} = \frac{1}{\zeta - a},$$

using (4).

$\exists$ constant c such that for w outside T, defined in (5), and for all n

$$\left| w - \frac{1}{n} \right| \geq c \cdot |w|.$$

By (5) it follows that

$$\left| h(f(\zeta)) - \frac{1}{n} \right| \geq c|h(f(\zeta))|, \qquad \zeta \in D,$$

or

$$\left| h(f(\zeta)) - \frac{1}{n} \right| \geq c|\zeta - a|\,|h_1(f(\zeta))|,$$

whence

$$|\phi_n(\zeta)| \leq \frac{1}{c}\frac{1}{|\zeta - a|}, \qquad \zeta \in D.$$

Thus (11) and (12) hold. Q.E.D.

Proof of Theorem 14.1. It suffices to show that $\mathfrak{A} = C(D)$.

Consider a measure μ on D orthogonal to $\mathfrak{A}$. We shall show that $\mu = 0$.

We know by Lemma 2.4 that $\int d|\mu|(z)/|\zeta - a| < \infty$ for almost all a in D. Fix such an a. Choose ϕ_n as in Lemma 14.3. By (12), then, $\phi_n \in L^1(|\mu|)$, and by (11) and (12), $\phi_n \to 1/\zeta - a$ pointwise on $D \setminus \{a\}$ and dominatedly with respect to $|\mu|$. Hence

$$\int \frac{d\mu(\zeta)}{\zeta - a} = \lim_{n \to \infty} \int \phi_n(\zeta)\, d\mu(\zeta).$$

Since $\phi_n \in \mathfrak{A}$ and $\mu \perp \mathfrak{A}$, the right-hand side is 0 for all n. Hence

$$\int \frac{d\mu(\zeta)}{\zeta - a} = 0 \qquad \text{for almost all } a \text{ in } D.$$

Also, since $\zeta \in \mathfrak{A}$, $1/\zeta - a \in \mathfrak{A}$ for $|a| > 1$ and so

$$\int \frac{d\mu(\zeta)}{\zeta - a} = 0 \qquad \text{for } |a| > 1.$$

Hence

$$\int \frac{d\mu(\zeta)}{\zeta - a} = 0 \qquad \text{for almost all } a \text{ in } \mathbf{C},$$

and so $\mu = 0$ by Lemma 2.7. We are done.

Note. We did not use the full strength of (3) in the proof, but only that (3) holds for almost all $\zeta_0 \in D$.

The natural question arises: What higher-dimensional analogues can be given for Theorem 14.1, with disks replaced by compact sets lying on smooth submanifolds of $\mathbf{C}^n$ whose real dimension is >2? To answer this question, entirely new techniques are required. Up to now we have been able to base all our applications of the $\bar{\partial}$-operator on the simple properties of that operator proved in Sections 6 and 7.

For the higher-dimensional generalizations of Theorem 14.1 we shall need a much deeper study of the $\bar{\partial}$-operator, made since 1958 by Morrey, Kohn, and Hörmander. We shall develop the necessary machinery in Section 16 and apply it in Section 17.

NOTES

Theorem 14.1 is due to the author, Polynomially convex disks, *Math. Ann.* **158** (1965). Only the case $n = 2$ is treated there, but the method is the same as that given here for general n. The use made of the Cauchy transform of a measure in the proof of the theorem goes back to E. Bishop's work on rational approximation in the plane. Theorem 14.1 was extended from disks to arbitrary smooth 2-manifolds by M. Freeman, Some conditions for uniform approximation on a manifold, *Function Algebras*, Scott, Foresman, Glenview, Ill., 1965.

15

THE FIRST COHOMOLOGY GROUP OF A MAXIMAL IDEAL SPACE

Given Banach algebras $\mathfrak{A}_1$ and $\mathfrak{A}_2$ with maximal ideal spaces $\mathscr{M}_1$ and $\mathscr{M}_2$, if $\mathfrak{A}_1$ and $\mathfrak{A}_2$ are isomorphic as algebras, then $\mathscr{M}_1$ and $\mathscr{M}_2$ are homeomorphic. It is thus to be expected that the topology of $\mathscr{M}(\mathfrak{A})$ is reflected in the algebraic structure of $\mathfrak{A}$, for an arbitrary Banach algebra $\mathfrak{A}$.

One result that we obtained in this direction was this: $\mathscr{M}$ is disconnected if and only if $\mathfrak{A}$ contains a nontrivial idempotent.

We now consider the first Čech cohomology group with integer coefficients, $H^1(\mathscr{M}, Z)$, of a maximal ideal space $\mathscr{M}$.

For decent topological spaces Čech cohomology coincides with singular or simplicial cohomology. We recall the definitions. Let X be a compact Hausdorff space. Fix an open covering $\mathscr{U} = \{U_\alpha\}$ of X, α running over some label set. We construct a simplicial complex as follows: Each U_α is a *vertex*, each pair (U_α, U_β) with $U_\alpha \cap U_\beta \neq \varnothing$ is a 1-*simplex*, and each triple $(U_\alpha, U_\beta, U_\gamma)$ with $U_\alpha \cap U_\beta \cap U_\gamma \neq \varnothing$ is a 2-*simplex*. A p-cochain ($p = 0, 1, 2$) is a map c^p assigning to each p-simplex an integer, and we require that c^p be an alternating function of its arguments; e.g., $c^1(U_\beta, U_\alpha) = -c^1(U_\alpha, U_\beta)$.

The totality of p-cochains forms a group under addition, denoted $C^p(\mathscr{U})$.

Define the coboundary $\delta : C^p(\mathscr{U}) \to C^{p+1}(\mathscr{U})$ as follows: For $c^0 \in C^0(\mathscr{U})$, (U_α, U_β) a 1-simplex,

$$\delta c^0(U_\alpha, U_\beta) = c^0(U_\beta) - c^0(U_\alpha).$$

For $c^1 \in C^1(\mathscr{U})$, $(U_\alpha, U_\beta, U_\gamma)$ a 2-simplex,

$$\delta c^1(U_\alpha, U_\beta, U_\gamma) = c^1(U_\alpha, U_\beta) + c^1(U_\beta, U_\gamma) + c^1(U_\gamma, U_\alpha).$$

c^1 is a 1-*cocycle* if $\delta c^1 = 0$. The set of all 1-cocycles forms a subgroup $\mathscr{Z}^1$ of $C^1(\mathscr{U})$, and $\delta C^0(\mathscr{U})$ is a subgroup of $\mathscr{Z}^1$. We define $H^1(\mathscr{U}, Z)$ as the quotient group

$\mathscr{Z}^1(\mathscr{U})/\delta C^0(\mathscr{U})$. We shall define the cohomology group $H^1(X, Z)$ as the "limit" of $H^1(\mathscr{U}, Z)$ as $\mathscr{U}$ gets finer and finer. More precisely

Definition 15.1. Given two coverings $\mathscr{U}$ and $\mathscr{V}$ of X, we say "$\mathscr{V}$ is finer than $\mathscr{U}$" ($\mathscr{V} > \mathscr{U}$) if for each V_α in $\mathscr{V}$ $\exists\phi(\alpha)$ in the label set of $\mathscr{U}$ with $V_\alpha \subset U_{\phi(\alpha)}$.

Note. ϕ is highly nonunique.

Under the relation $>$ the family $\mathscr{F}$ of all coverings of X is a directed set. We have a map

$$\mathscr{U} \to H^1(\mathscr{U}, Z)$$

of this directed set to the family of groups $H^1(\mathscr{U}, Z)$.

For a discussion of direct systems of groups and their application to cohomology we refer the reader to W. Hurewicz and H. Wallman, *Dimension Theory* (Princeton University Press, Princeton, N.J., 1948, Chap. 8, Sec. 4) and shall denote this reference by H.-W.

To each pair $\mathscr{U}$ and $\mathscr{V}$ of coverings of X with $\mathscr{V} > \mathscr{U}$ corresponds for each p a map ρ:

$$C^p(\mathscr{U}) \to C^p(\mathscr{V}),$$

where $\rho c^p(V_{\alpha_0}, V_{\alpha_1}, \ldots, V_{\alpha_p}) = c^p(U_{\phi(\alpha_0)}, \ldots, U_{\phi(\alpha_p)})$, ϕ being as in Definition 15.1.

LEMMA 15.1

ρ induces a homomorphism $K^{\mathscr{U},\mathscr{V}}: H^p(\mathscr{U}, Z) \to H^p(\mathscr{V}, Z)$.

LEMMA 15.2

$K^{\mathscr{U},\mathscr{V}}$ depends only on $\mathscr{U}$ and $\mathscr{V}$, not on the choice of ϕ.

For the proofs see H.-W.

The homomorphisms $K^{\mathscr{U},\mathscr{V}}$ make the family $\{H^p(\mathscr{U}, Z)|\mathscr{U}\}$ into a direct system of groups.

Definition 15.2. $H^1(X, Z)$ is the limit group of the direct system of groups $\{H^1(\mathscr{U}, Z)|\mathscr{U}\}$.

$\exists$ a homomorphism $K^{\mathscr{U}}: H^1(\mathscr{U}, Z) \to H^1(X, Z)$ such that for $\mathscr{V} > \mathscr{U}$ we have

$$K^{\mathscr{V}} \circ K^{\mathscr{U},\mathscr{V}} = K^{\mathscr{U}}. \tag{1}$$

(See H.-W.)

Our goal is the following result: Let $\mathfrak{A}$ be a Banach algebra. Put

$$\mathfrak{A}^{-1} = \{x \in \mathfrak{A}|x \text{ has an inverse in } \mathfrak{A}\}$$

and

$$\exp \mathfrak{A} = \{x \in \mathfrak{A}|x = e^y \text{ for some } y \in \mathfrak{A}\}.$$

$\mathfrak{A}^{-1}$ is a group under multiplication and $\exp \mathfrak{A}$ is a subgroup of $\mathfrak{A}^{-1}$.

THEOREM 15.3 (ARENS–ROYDEN)

Let $\mathscr{M} = \mathscr{M}(\mathfrak{A})$. Then $H^1(\mathscr{M}, Z)$ is isomorphic to the quotient group $\mathfrak{A}^{-1}/\exp \mathfrak{A}$.

COROLLARY

If $H^1(\mathcal{M}, Z) = 0$, then every invertible element x in $\mathfrak{A}$ admits a representation $x = e^y$, $y \in \mathfrak{A}$.

Exercise 15.1. Let $\mathfrak{A} = C(\Gamma)$, Γ the circle. Verify Theorem 15.3 in this case.

Exercise 15.2. Do the same for $\mathfrak{A} = C(I)$, I the unit interval.

In the exercises, take as given that $H^1(\Gamma, Z) = Z$ and $H^1(I, Z) = \{0\}$.

THEOREM 15.4

Let X be a compact space. $\exists$ a natural homomorphism

$$\eta : C(X)^{-1} \to H^1(X, Z)$$

such that η is onto and the kernel of η $=$ exp $C(X)$.

Proof. Fix $f \in C(X)^{-1}$. Thus $f \neq 0$ on X. We shall associate to f an element of $H^1(X, Z)$, to be denoted $\eta(f)$.

Let $\mathcal{U} = \{U_\alpha\}$ be an open covering of X. A set of functions $g_\alpha \in C(U_\alpha)$ will be called *$(f, \mathcal{U})$-admissible* if

$$f = e^{g_\alpha} \text{ in } U_\alpha \tag{2}$$

and

$$|g_\alpha(x) - g_\alpha(y)| < \pi \qquad \text{for } x,y \text{ in } U_\alpha. \tag{3}$$

Such admissible sets exist whenever $f(U_\alpha)$ lies, for each α, in a small disk excluding 0. Equations (2) and (3) imply that $g_\beta - g_\alpha$ is constant in $U_\alpha \cap U_\beta$.

Now fix a covering $\mathcal{U}$ and an $(f, \mathcal{U})$-admissible set g_α. Then $\exists$ integers $h_{\alpha\beta}$ with

$$\frac{1}{2\pi i}(g_\beta - g_\alpha) = h_{\alpha\beta} \text{ in } U_\alpha \cap U_\beta.$$

The map $h : (U_\alpha, U_\beta) \to h_{\alpha\beta}$ is an element of $C^1(\mathcal{U})$; in fact, h is a 1-cocycle. For given any 1-simplex $(U_\alpha, U_\beta, U_\gamma)$

$$\delta h(U_\alpha, U_\beta, U_\gamma) = h_{\alpha\beta} + h_{\beta\gamma} + h_{\gamma\alpha}$$

$$= \frac{1}{2\pi i}\{g_\beta - g_\alpha + g_\gamma - g_\beta + g_\alpha - g_\gamma\} = 0$$

at each point of $U_\alpha \cap U_\beta \cap U_\gamma$.

Denote by $[h]$ the cohomology class of h in $H^1(\mathcal{U}, Z)$.

(4) $[h]$ is independent of the choice of $\{g_\alpha\}$ and depends only on f and $\mathcal{U}$.

For let $\{g'_\alpha\}$ be another $(f, \mathcal{U})$-admissible set. By (2) and (3), $\exists k_\alpha \in Z$ with

$$g'_\alpha(x) - g_\alpha(x) = 2\pi i k_\alpha \qquad \text{for } x \in U_\alpha.$$

The cocycle h' determined by $\{g'_\alpha\}$ is given by

$$h'_{\alpha\beta} = h'(U_\alpha, U_\beta) = \frac{1}{2\pi i}(g'_\beta(x) - g'_\alpha(x))$$

$(x \in U_\alpha \cap U_\beta)$. Hence

$$h'_{\alpha\beta} = h_{\alpha\beta} + \delta k,$$

where k is the 0-cochain in $C^0(\mathscr{U})$ defined by $k(U_\alpha) = k_\alpha$. Thus $[h'] = [h]$, as desired.

We define

$$\eta_{\mathscr{U}}(f) = [h]$$

and

$$\eta(f) = K^{\mathscr{U}}([h]) \in H^1(X, Z).$$

Using (1) we can verify that $\eta(f)$ depends only on f, not on the choice of the covering $\mathscr{U}$.

(5) η maps $C(X)^{-1}$ onto $H^1(X, Z)$.

To prove this fix $\xi \in H^1(X, Z)$. Choose a covering $\mathscr{U}$ and a cocycle h in $C^1(\mathscr{U})$ with $K^{\mathscr{U}}([h]) = \xi$. Put $h_{\nu\mu} = h(U_\nu, U_\mu)$. Since X is compact and so an arbitrary open covering admits a finite covering finer than itself, we may assume that $\mathscr{U}$ is finite, $\mathscr{U} = \{U_1, U_2, \ldots, U_s\}$.

Choose a partition of unity χ_α, $1 \le \alpha \le s$, with supp $\chi_\alpha \subset U_\alpha$, $\chi_\alpha \in C(X)$, and $\sum_{\alpha=1}^{s} \chi_\alpha = 1$. For each k define

$$g_k = 2\pi i \sum_{\nu=1}^{s} h_{\nu k}\chi_\nu(x) \qquad \text{for } x \in U_k,$$

where we put $h_{\nu k} = 0$ unless U_ν meets U_k. Then $g_k \in C(U_k)$. Fix $x \in U_j \cap U_k$. Note that unless U_ν meets $U_k \cap U_j$, $\chi_\nu(x) = 0$. Then

$$(g_k - g_j)(x) = 2\pi i \sum_\nu \chi_\nu(x)(h_{\nu k} - h_{\nu j}).$$

Since h is a 1-cocycle, $h_{k\nu} + h_{\nu j} + h_{jk} = 0$ whenever $U_k \cap U_\nu \cap U_j \neq \varnothing$. Hence in $U_j \cap U_k$,

$$g_k - g_j = 2\pi i \sum_\nu \chi_\nu h_{jk} = 2\pi i h_{jk}.$$

Define f_α in U_α by $f_\alpha = e^{g_\alpha}$. Then $f_\alpha \in C(U_\alpha)$ and in $U_\alpha \cap U_\beta$,

$$\frac{f_\beta}{f_\alpha} = e^{g_\beta - g_\alpha} = e^{2\pi i h_{\alpha\beta}} = 1.$$

Thus $f_\alpha = f_\beta$ in $U_\alpha \cap U_\beta$, so the different f_α fit together to a single function f in $C(X)$. Also,

$$f_\alpha = e^{g_\alpha} \text{ in } U_\alpha \qquad \text{and} \qquad g_\beta - g_\alpha = 2\pi i h_{\alpha\beta} \text{ in } U_\alpha \cap U_\beta.$$

From this and the definition of η, we can verify that $\eta(f) = K^{\mathscr{U}}([h]) = \xi$.

(6) Fix f in the kernel of η. Then $f \in \exp C(X)$.

For $\eta(f)$ is the zero element of $H^1(X, Z)$. Hence $\exists$ covering $\mathscr{V}$ such that if h is the cocycle in $C^1(\mathscr{V})$ associated to f by our construction, then the cohomology class of h is 0; i.e., if

$$f = e^{g_\alpha} \text{ in } U_\alpha,$$

then $\exists H \in C^0(\mathscr{V})$ such that

$$g_\beta - g_\alpha = 2\pi i(H_\beta - H_\alpha) \text{ in } V_\alpha \cap V_\beta.$$

Then

$$g_\beta - 2\pi i H_\beta = g_\alpha - 2\pi i H_\alpha \text{ in } V_\alpha \cap V_\beta.$$

Hence $\exists$ global function G in $C(X)$ with $G = g_\alpha - 2\pi i H_\alpha$ in V_α for each α. Then $f = e^G$, and we are done.

Since it is clear that η vanishes on $\exp C(X)$, the proof of Theorem 15.4 is complete.

Note. We leave to the reader to verify that η is natural.

Now let X be a compact space and $\mathscr{L}$ a subalgebra of $C(X)$. The map η (of Theorem 15.4) restricts to $\mathscr{L}^{-1} = \{f \in \mathscr{L} | 1/f \in \mathscr{L}\}$, mapping $\mathscr{L}^{-1}$ into $H^1(X, Z)$.

Definition 15.3. $\mathscr{L}$ is *full* if

(a) η maps $\mathscr{L}^{-1}$ onto $H^1(X, Z)$.

(b) $x \in \mathscr{L}^{-1}$ and $\eta(x) = 0$ imply $\exists y \in \mathscr{L}$, with $x = e^y$.

Next let X be a compact polynomially convex subset of $\mathbf{C}^n$.

Definition 15.4. $\mathscr{H}(X) = \{f \in C(X) | \exists$ neighborhood U of X and $\exists F \in H(U)$ with $F = f$ on $X\}$.

$\mathscr{H}(X)$ is a subalgebra of $C(X)$.

LEMMA 15.5

$\mathscr{H}(X)$ is full.

Proof. Fix $\gamma \in H^1(X, Z)$. Then $\exists$ a covering $\mathscr{U}$ of X and a cocycle $h \in C^1(\mathscr{U})$ with $K^{\mathscr{U}}([h]) = \gamma$.

Without loss of generality, we may assume that

$$\mathscr{U} = \{U_\alpha \cap X | 1 \le \alpha \le s\}, \text{ each } U_\alpha \text{ open in } \mathbf{C}^n.$$

(Why?)

For each α choose $\chi_\alpha \in C_0^\infty(U_\alpha)$, with $\Sigma_{\alpha=1}^s \chi_\alpha = 1$ in some neighborhood N of X. Put $h_{\alpha\beta} = h(U_\alpha \cap X, U_\beta \cap X) \in Z$. Fix α and put for $x \in U_\alpha$,

$$g_\alpha(x) = 2\pi i \sum_{\nu=1}^{s} h_{\nu\alpha}\chi_\nu(x),$$

where $h_{\nu\alpha} = 0$ unless $U_\nu \cap U_\alpha \neq \varnothing$. Then $g_\alpha \in C^\infty(U_\alpha)$, and, as in the proof of Theorem 15.4, we have in $U_\alpha \cap U_\beta \cap N$,

(7) $$g_\beta - g_\alpha = 2\pi i h_{\alpha\beta}.$$

Hence $\bar{\partial} g_\beta - \bar{\partial} g_\alpha = 0$ in $U_\alpha \cap U_\beta \cap N$, so the $\bar{\partial} g_\alpha$ fit together to a $\bar{\partial}$-closed (0, 1)-form defined in N.

By Lemma 7.4 $\exists$ a p-polyhedron Π with $X \subset \Pi \subset N$. By Theorem 7.6 $\exists$ a neighborhood W of Π and $u \in C^\infty(W)$ with

$$\bar{\partial} u = \bar{\partial} g_\alpha \text{ in } W \cap U_\alpha. \tag{8}$$

Put $V_\alpha = U_\alpha \cap W$, $1 \leq \alpha \leq s$. Then $\mathfrak{A} = \{V_\alpha \cap X | 1 \leq \alpha \leq s\}$.

Put $g'_\alpha = g_\alpha - u$ in V_α. Then $g'_\alpha \in H(V_\alpha)$, by (8). Also, by (7),

$$\frac{1}{2\pi i}(g'_\beta - g'_\alpha) = \frac{1}{2\pi i}(g_\beta - g_\alpha) = h_{\alpha\beta} \text{ in } V_\alpha \cap V_\beta.$$

Define $f = e^{g'_\alpha}$ in V_α for each α. In $V_\alpha \cap V_\beta$ the two definitions of f are

$$e^{g'_\alpha} \text{ and } e^{g'_\beta} = e^{g'_\alpha + 2\pi i h_{\alpha\beta}} = e^{g'_\alpha}.$$

Hence f is well defined in $\bigcup_\alpha V_\alpha$ and holomorphic there, so $f|_X \in \mathscr{H}(X)$ and, in fact, $\in (\mathscr{H}(X))^{-1}$.

$g'_\beta - g'_\alpha = 2\pi i h_{\alpha\beta}$, whence $\eta(f) = K^{\mathscr{U}}([h]) = \gamma$. We have verified (a) in Definition 15.3.

Now fix $f \in (\mathscr{H}(X))^{-1}$ with $\eta(f) = 0$. Let F be holomorphic in a neighborhood of X with $F = f$ on X.

Since $\eta(f) = 0$, $\eta_{\mathscr{U}}(f) = 0$ for some covering $\mathscr{U}$. Choose a covering of X by open subsets W_α of $\mathbf{C}^n$, $1 \leq \alpha \leq s$, such that

$$\mathscr{W} > \mathscr{U}. \tag{9}$$

$$\exists G_\alpha \in H(W_\alpha) \qquad \text{with } F = e^{G_\alpha} \text{ in } W_\alpha. \tag{10}$$

$$|G_\alpha(x) - G_\alpha(y)| < \pi \qquad \text{for } x, y \in W_\alpha. \tag{11}$$

$$\text{If } W_\alpha \cap W_\beta \neq \varnothing, \text{ then } W_\alpha \cap W_\beta \text{ meets } X. \tag{12}$$

Let $\mathscr{W} = \{W_\alpha \cap X | 1 \leq \alpha \leq s\}$. $\eta_{\mathscr{U}}(f) = 0$, so $\eta_{\mathscr{W}}(f) = 0$. Hence $\exists$ integers k_α such that if $(W_\alpha \cap X) \cap (W_\beta \cap X) \neq \varnothing$, then in $(W_\alpha \cap X) \cap (W_\beta \cap X)$,

$$\frac{1}{2\pi i}(G_\beta - G_\alpha) = k_\beta - k_\alpha. \tag{13}$$

Now fix α and β with $W_\alpha \cap W_\beta \neq \varnothing$. By (12), $(W_\alpha \cap X) \cap (W_\beta \cap X) \neq \varnothing$. Hence, by (13),

$$G_\beta - G_\alpha = 2\pi i k_\beta - 2\pi i k_\alpha \text{ in } W_\alpha \cap W_\beta \cap X.$$

Also, because of (10) and (11),

$$G_\beta - G_\alpha \text{ is constant in } W_\alpha \cap W_\beta.$$

Hence

$$G_\beta - G_\alpha = 2\pi i k_\beta - 2\pi i k_\alpha \text{ in } W_\alpha \cap W_\beta$$

or

$$G_\beta - 2\pi i k_\beta = G_\alpha - 2\pi i k_\alpha \text{ in } W_\alpha \cap W_\beta.$$

Hence $\exists G \in H(\bigcup_\alpha W_\alpha)$ with $G = G_\alpha - 2\pi i k_\alpha$ in W_α for each α. Then

$$F = e^G \text{ in } \bigcup_\alpha W_\alpha.$$

Since $G|_X \in \mathscr{H}(X)$, we have verified (b) in Definition 15.3. So the lemma is proved.

LEMMA 15.6

Let $\mathscr{L}$ be a finitely generated uniform algebra on a space X with $X = \mathscr{M}(\mathscr{L})$. Then $\mathscr{L}$ is full [as subalgebra of $C(X)$.]

Proof. By Exercise 7.3 it suffices to assume that $\mathscr{L} = P(X)$, X a compact polynomially convex set in $\mathbf{C}^n$.

By the Oka–Weil theorem $\mathscr{H}(X) \subset P(X)$. Fix $\gamma \in H^1(X, Z)$. By the last lemma, $\exists f \in (\mathscr{H}(X))^{-1}$ with $\eta(f) = \gamma$. Then $f \in (P(X))^{-1}$. Thus η maps $(P(X))^{-1}$ onto $H^1(X, Z)$. Now fix $f \in (P(X))^{-1}$ with $\eta(f) = 0$, and fix $\varepsilon > 0$. Choose a polynomial g with

$$\|g - f\| < \varepsilon < \inf_X |f|,$$

the norm being taken in $P(X)$. Put $h = (f - g)/f$. Then $\|h\| < 1$ and $g = f(1 - h)$. Hence $1 - h \in \exp C(X)$ (why?) and so $\eta(1 - h) = 0$. Hence

$$\eta(g) = \eta(f) = 0.$$

But $g \in (\mathscr{H}(X))^{-1}$, whence by the last lemma $\exists g^* \in \mathscr{H}(X)$ with $g = e^{g^*}$.

Also, $1 - h = e^k$ for some $k \in P(X)$, since $\|h\| < 1$. (Why?) Hence $f = e^{g^* - k}$, so $f \in \exp(P(X))$. Thus $P(X)$ is full. Q.E.D.

To extend this result to a uniform algebra $\mathfrak{A}$ that fails to be finitely generated, we may express $\mathfrak{A}$ as a "limit" of its finitely generated subalgebras. For this extension we refer the reader to H. Royden, Function algebras, *Bull. Am. Math. Soc.* **69** (1963), 281–298. The following is proved there (Proposition 11):

LEMMA 15.7

Let $\mathscr{L}$ be an arbitrary uniform algebra on a space X with $X = \mathscr{M}(\mathscr{L})$. Then $\mathscr{L}$ is full.

Proof of Theorem 15.3. Put $X = \mathscr{M}$ and let $\mathscr{L}$ be the uniform closure of $\hat{\mathfrak{A}}$ on X. Then $X = \mathscr{M}(\mathscr{L})$. By Lemma 15.7 $\mathscr{L}$ is full [as subalgebra of $C(X)$].

Let $x \in \mathfrak{A}^{-1}$. Then $\hat{x} \in (C(X))^{-1}$. Define a map Φ of $\mathfrak{A}^{-1}$ into $H^1(X, Z)$ by

$$\Phi(x) = \eta(\hat{x}).$$

We claim Φ is onto $H^1(X, Z)$. Fix $\gamma \in H^1(X, Z)$. Since $\mathscr{L}$ is full, $\exists f \in \mathscr{L}^{-1}$ with $\eta(f) = \gamma$. Choose $\varepsilon > 0$ with $\inf_X|f| > \varepsilon$, and choose $g \in \mathfrak{A}$ with $|\hat{g} - f| < \varepsilon$ on X. Then $g \in \mathfrak{A}^{-1}$, $\hat{g} = f(1 - (f - \hat{g})/f)$, and $\sup_X|(f - \hat{g})/f| < 1$. Hence $\exists b \in C(X)$

with $1 - (f - \hat{g})/f = e^b$, and so $\eta(\hat{g}) = \eta(f) = \gamma$. Thus $\Phi(g) = \gamma$, so Φ is onto, as claimed.

Next we claim that the kernel of $\Phi = \exp \mathfrak{A}$. Since one direction is clear, it remains to show that $x \in \mathfrak{A}^{-1}$ and $\Phi(x) = 0$ implies that $x \in \exp \mathfrak{A}$.

Then fix $x \in \mathfrak{A}^{-1}$ with $\Phi(x) = \eta(\hat{x}) = 0$. Since $\mathscr{L}$ is full and $\hat{x} \in \mathscr{L}^{-1}$, $\exists F \in \mathscr{L}$ with $\hat{x} = e^F$. Since F is in the uniform closure of $\hat{\mathfrak{A}}$, e^F is in the uniform closure of functions $e^{\hat{h}}$, $h \in \mathfrak{A}$.

Hence $\exists g = e^h$ with $h \in \mathfrak{A}$ and

$$|\hat{x} - \hat{g}| < \tfrac{1}{3} \inf_X |\hat{x}| \text{ on } X.$$

Then

$$|\hat{g}| > \tfrac{2}{3} \inf_X |\hat{x}|, \qquad \text{so } \frac{1}{|\hat{g}|} < \frac{3}{2} \cdot \frac{1}{\inf_X |\hat{x}|}.$$

Hence uniformly on X,

$$|1 - \hat{x}\hat{g}^{-1}| = |\hat{x} - \hat{g}| \cdot |\hat{g}^{-1}| < \tfrac{1}{2}.$$

It follows that for large n, $\|(1 - xg^{-1})^n\|^{1/n} < \frac{3}{4}$, and so the series

$$-\sum_1^\infty \frac{1}{n}(1 - xg^{-1})^n$$

converges in $\mathfrak{A}$ to an element k. Since

$$\log(1 - z) = -\sum_1^\infty \frac{1}{n} z^n, \qquad |z| < 1,$$

$k = \log(xg^{-1})$, so that $xg^{-1} = e^k$. Hence $x = e^{k+h} \in \exp \mathfrak{A}$. Hence the kernel of Φ is $\exp \mathfrak{A}$, as claimed.

Φ thus induces an isomorphism of $\mathfrak{A}^{-1}/\exp \mathfrak{A}$ onto $H^1(X, Z)$, and Theorem 15.3 is proved.

Note. No analogous algebraic interpretation of the higher cohomology groups $H^p(\mathscr{M}, Z)$, $p > 1$, has so far been obtained. However, one has the following result:

THEOREM 15.8

Let $\mathfrak{A}$ be a Banach algebra with n generators. Then $H^p(\mathscr{M}, \mathbf{C}) = 0$, $p \geq n$.

This result is due to A. Browder, Cohomology of maximal ideal spaces, *Bull. Am. Math. Soc.* **67** (1961), 515–516. Observe that if $\mathfrak{A}$ has n generators, then $\mathscr{M}$ is homeomorphic to a subset of $\mathbf{C}^n$ and hence that the vanishing of $H^p(\mathscr{M}, \mathbf{C})$ is obvious for $p \geq 2n$.

NOTES

For the first theorem of the type studied in this section (Theorem 15.4) see S. Eilenberg, Transformations continues en circonférence et la topologie du plan, *Fund.*

Math. **26** (1936) and N. Bruschlinsky, Stetige Abbildungen und Bettische Gruppen der Dimensionszahl 1 und 3, *Math. Ann.* **109** (1934). Theorem 15.3 is due to R. Arens, The group of invertible elements of a commutative Banach algebra, *Studia Math.* **1** (1963), and H. Royden, Function algebras, *Bull. Am. Math. Soc.* **69** (1963). The proof we have given follows Royden's paper.

16

THE $\bar{\partial}$-OPERATOR IN SMOOTHLY BOUNDED DOMAINS

Let Ω be a bounded open subset of $\mathbf{C}^n$. We are essentially concerned with the following problem: Given a form f of type (0, 1) on Ω with $\bar{\partial}f = 0$, find a function u on Ω such that $\bar{\partial}u = f$.

In order to be able to use the properties of operators on Hilbert space in attacking this question, we shall consider L^2-spaces rather than (as before) spaces of smooth functions.

$L^2(\Omega)$ denotes the space of measurable functions u on Ω with $\int_\Omega |u|^2\, dV < \infty$, where dV is Lebesgue measure.

$L^2_{0,1}(\Omega)$ is the space of (0, 1)-forms

$$f = \sum_{j=1}^{n} f_j\, d\bar{z}_j,$$

where each $f_j \in L^2(\Omega)$. Put $|f|^2 = \Sigma_{j=1}^n |f_j|^2$. Analogously, $L^2_{0,2}(\Omega)$ is the space of (0, 2)-forms

$$\phi = \sum_{i<j} \phi_{ij}\, d\bar{z}_i \wedge d\bar{z}_j,$$

where each $\phi_{ij} \in L^2(\Omega)$.

We shall define an operator T_0 from a subspace of $L^2(\Omega)$ to $L^2_{0,1}(\Omega)$ such that T_0 coincides with $\bar{\partial}$ on functions that are smooth on $\bar{\Omega}$.

Definition 16.1. Let $u \in L^2(\Omega)$. Fix $k \in L^2(\Omega)$ and fix j, $1 \leq j \leq n$. We say

$$\frac{\partial u}{\partial \bar{z}_j} = k$$

if for all $g \in C_0^\infty(\Omega)$ we have

$$-\int_\Omega u \frac{\partial g}{\partial \bar{z}_j}\, dV = \int_\Omega gk\, dV.$$

Note. Thus $k = \partial u/\partial \bar{z}_j$ in the sense of the theory of distributions. If u is smooth on $\bar{\Omega}$, then $k = \partial u/\partial \bar{z}_j$ in the usual sense.

Definition 16.2

$$\mathscr{D}_{T_0} = \left\{ u \in L^2(\Omega) \mid \text{for each } j,\ 1 \le j \le n\ \exists k_j \in L^2(\Omega) \text{ with } \frac{\partial u}{\partial \bar{z}_j} = k_j \right\}.$$

For $u \in \mathscr{D}_{T_0}$,

$$T_0 u = \sum_{j=1}^{n} \frac{\partial u}{\partial \bar{z}_j} d\bar{z}_j \in L^2_{0,1}(\Omega).$$

Fix $\Sigma_{j=1}^n f_j\, d\bar{z}_j$ with each $f_j \in L^2(\Omega)$. $\partial f_j/\partial \bar{z}_k$ and $\partial f_k/\partial \bar{z}_j$ are defined as distributions.

Definition 16.3

$$\mathscr{D}_{S_0} = \left\{ f = \sum_{j=1}^{n} f_j\, d\bar{z}_j \in L^2_{0,1} \middle| \frac{\partial f_j}{\partial \bar{z}_k} - \frac{\partial f_k}{\partial \bar{z}_j} \in L^2(\Omega), \text{ all } j,k \right\}.$$

For $f \in \mathscr{D}_{S_0}$,

$$S_0 f = \sum_{j<k} \left(\frac{\partial f_j}{\partial \bar{z}_k} - \frac{\partial f_k}{\partial \bar{z}_j} \right) d\bar{z}_k \wedge d\bar{z}_j \in L^2_{0,2}(\Omega).$$

Note that S_0 coincides with $\bar{\partial}$ on smooth forms f. Note also that if $u \in \mathscr{D}_{T_0}$, then $T_0 u \in \mathscr{D}_{S_0}$, and

(1) $$S_0 \cdot T_0 = 0.$$

Now let Ω be defined by the inequality $\rho < 0$, where ρ is a smooth real-valued function in some neighborhood of $\bar{\Omega}$. Assume that the gradient of $\rho \neq 0$ on $\partial\Omega$. We impose on ρ the following condition:

(2) For all $z \in \partial\Omega$, if $(\xi_1, \ldots, \xi_n) \in \mathbf{C}^n$ and $\sum_j \partial\rho/\partial z_j(z)\xi_j = 0$,

then

$$\sum_{j,k} \frac{\partial^2 \rho}{\partial z_j\, \partial \bar{z}_k}(z)\xi_j \bar{\xi}_k \ge 0.$$

THEOREM 16.1

Let ρ satisfy condition (2). *For every $g \in L^2_{0,1}(\Omega)$ with $S_0 g = 0$, $\exists u \in \mathscr{D}_{T_0}$ such that*

(a) $T_0 u = g$, *and*

(b) $$\int_\Omega |u|^2\, dV \le e^{R^2} \cdot \int_\Omega |g|^2\, dV,$$

if $\Omega \subset \{z \in \mathbf{C}^n \mid |z| \le R\}$.

We need some general results about linear operators on Hilbert space.

Let H_1 and H_2 be Hilbert spaces, and let A be a linear transformation from a dense subspace $\mathscr{D}_A$ of H_1 into H_2.

Definition 16.4. A is *closed* if for each sequence $g_n \in \mathscr{D}_A$,

$$g_n \to g \quad \text{and} \quad Ag_n \to h$$

implies that $g \in \mathscr{D}_A$ and $Ag = h$.

Definition 16.5

$$\mathscr{D}_{A^*} = \{x \in H_2 | \exists x^* \in H_1 \text{ with } (Au, x) = (u, x^*) \text{ for all } u \in \mathscr{D}_A.\}$$

Since $\mathscr{D}_A$ is dense, x^* is unique if it exists. For $x \in \mathscr{D}_{A^*}$, define $A^*x = x^*$. A^* is called the *adjoint* of A. $\mathscr{D}_{A^*}$ is a linear space and A^* is a linear transformation of $\mathscr{D}_{A^*} \to H_1$.

PROPOSITION

If A is closed, then $\mathscr{D}_{A^*}$ is dense in H_2. Moreover, if $\beta \in H_1$ and if for some constant δ

$$|(A^*f, \beta)| \leq \delta \|f\|$$

for all $f \in \mathscr{D}_{A^*}$, then $\beta \in \mathscr{D}_A$.

For the proof of this proposition and related matters the reader may consult, e.g., F. Riesz and B. Sz.-Nagy, *Leçons d'analyse fonctionelle*, Budapest, 1953, Chap. 8.

Consider now three Hilbert spaces H_1, H_2, and H_3 and densely defined and closed linear operators

$$T : H_1 \to H_2 \quad \text{and} \quad S : H_2 \to H_3.$$

Assume that

(3) $$S \cdot T = 0;$$

i.e., for $f \in \mathscr{D}_T$, $Tf \in \mathscr{D}_S$ and $S(Tf) = 0$.

We write $(u, v)_j$ for the inner product of u and v in H_j, $j = 1, 2, 3$, and similarly $\|u\|_j$ for the norm in H_j.

THEOREM 16.2

Assume $\exists$ a constant c such that for all $f \in \mathscr{D}_{T^*} \cap \mathscr{D}_S$,

(*) $$\|T^*f\|_1^2 + \|Sf\|_3^2 \geq c^2 \|f\|_2^2.$$

Then if $g \in H_2$ with $Sg = 0$, $\exists u \in \mathscr{D}_T$ such that

(4) $$Tu = g$$

and

(5) $$\|u\|_1 \leq \frac{1}{c} \|g\|_2.$$

Proof. Put $N_S = \{h \in \mathscr{D}_S | Sh = 0\}$. N_S is a closed subspace of H_2. (Why?)

We claim that if $g \in N_S$, then

$$(6) \qquad |(g, f)_2| \leq \frac{1}{c}\|T^*f\|_1 \cdot \|g\|_2,$$

for all $f \in \mathscr{D}_{T^*}$.

To show this, fix $f \in \mathscr{D}_{T^*}$.

$$f = f' + f'', \qquad \text{where } f' \perp N_S, f'' \in N_S.$$

By (*) we have $\|T^*f''\|_1 \geq c\|f''\|_2$. Then

$$|(f, g)_2| = |(f'', g)_2| \leq \|g\|_2 \cdot \|f''\|_2 \leq \frac{1}{c}\|g\|_2 \cdot \|T^*f''\|_1.$$

But $T^*f' = 0$, for if $h \in \mathscr{D}_T$, $(Th, f') = (h, T^*f')$ and the left-hand side $= 0$, because $f' \perp N_S$ while $S(Th) = 0$ by (3). Hence $T^*f = T^*f''$, and so (6) holds, as claimed.

We now define a linear functional L on the range of T^* in H_1 by

$$L(T^*f) = (f, g)_2, \qquad f \in \mathscr{D}_{T^*}, g \text{ fixed in } N_S.$$

By (6), then,

$$|L(T^*f)| \leq \frac{1}{c}\|g\|_2\|T^*f\|_1.$$

It follows that L is well defined on the range of T^* and that $\|L\| \leq (1/c)\|g\|_2$. Hence $\exists u \in H_1$ representing L; i.e.,

$$L(T^*f) = (T^*f, u)_1,$$

and $\|u\|_1 = \|L\|$. It follows by the proposition that $u \in \mathscr{D}_T$, and

$$(f, g)_2 = (T^*f, u)_1 = (f, Tu)_2,$$

all $f \in \mathscr{D}_{T^*}$.

Hence $g = Tu$, and $\|u\|_1 \leq (1/c)\|g\|_2$. Thus (4) and (5) are established. Q.E.D.

It is now our task to verify hypothesis (*) for our operators T_0 and S_0 in order to apply Theorem 16.2 to the proof of Theorem 16.1. This means that we must find a lower bound for $\|T_0^*f\|^2 + \|S_0f\|^2$. For this purpose it is advantageous to use not the usual inner product on $L^2(\Omega)$ but an equivalent inner product based on a weight function.

Let ϕ be a smooth positive function defined in a neighborhood of $\bar{\Omega}$. Put $H_1 = L^2(\Omega)$ with the inner product

$$(f, g)_1 = \int_\Omega f\bar{g}e^{-\phi}\, dV.$$

Similarly, let H_2 be the Hilbert space obtained by imposing on $L^2_{0,1}(\Omega)$ the inner product

$$\left(\sum_{j=1}^n f_j\, d\bar{z}_j, \sum_{j=1}^n g_j\, d\bar{z}_j\right)_2 = \int_\Omega \left(\sum_{j=1}^n f_j\bar{g}_j\right)e^{-\phi}\, dV.$$

Finally define H_3 in an analogous way by putting a new inner product on $L^2_{0,2}(\Omega)$. Then

$$T_0 : H_1 \to H_2, \qquad S_0 : H_2 \to H_3.$$

It is easy to verify that $\mathscr{D}_{T_0}, \mathscr{D}_{S_0}$ are dense subspaces of H_1 and H_2, respectively, and that T_0 and S_0 are closed operators. Our basic result is the following: Define $C^1_{0,1}(\bar{\Omega}) = \{f = \sum_{j=1}^n f_j\, d\bar{z}_j |$ each $f_j \in C^1$ in a neighborhood of $\bar{\Omega}.\}$

THEOREM 16.3

Fix f in $C^1_{0,1}(\bar{\Omega})$. Let $f \in \mathscr{D}_{T_0^*} \cap \mathscr{D}_{S_0}$. Then

$$(7) \qquad \|T_0^* f\|_1^2 + \|S_0 f\|_3^2 = \sum_{j,k} \int_\Omega f_j \bar{f}_k \frac{\partial^2 \phi}{\partial z_j\, \partial \bar{z}_k} e^{-\phi}\, dV$$

$$+ \sum_{j,k} \int_\Omega \left|\frac{\partial f_k}{\partial \bar{z}_j}\right|^2 e^{-\phi}\, dV + \sum_{j,k} \int_{\partial\Omega} f_j \bar{f}_k \frac{\partial^2 \rho}{\partial z_j\, \partial \bar{z}_k} e^{-\phi}\, dS,$$

dS denoting the element of surface area on $\partial\Omega$.

Suppose for the moment that Theorem 16.3 has been established. Put

$$\phi(z) = \sum_{j=1}^n |z_j|^2 = |z|^2.$$

Then $\partial^2\phi/\partial z_j\, \partial \bar{z}_k = 0$ if $j \neq k$, $= 1$ if $j = k$. The first integral on the right in (7) is now

$$\sum_{j=1}^n \int_\Omega |f_j|^2 e^{-\phi}\, dV = \|f\|_2^2.$$

The second integral is evidently ≥ 0. Now

$$\sum_{j,k} \frac{\partial^2 \rho}{\partial z_j\, \partial \bar{z}_k} f_j \bar{f}_k \geq 0 \qquad \text{if} \sum_j \frac{\partial \rho}{\partial z_j} f_j = 0 \text{ on } \partial\Omega,$$

by (2). Hence (7) gives

$$(8) \qquad \|T_0^* f\|_1^2 + \|S_0 f\|_3^2 \geq \|f\|_2^2,$$

if

$$(9) \qquad \sum_j \frac{\partial \rho}{\partial z_j} f_j = 0 \text{ on } \partial\Omega.$$

We shall show below that (9) holds whenever $f \in \mathscr{D}_{T_0^*} \cap \mathscr{D}_{S_0}$ and f is C^1 in a neighborhood of $\bar{\Omega}$. Thus Theorem 16.3 implies that (8) holds for each smooth f in $\mathscr{D}_{T_0^*} \cap \mathscr{D}_{S_0}$.

We now quote a result from the theory of partial differential operators, which seems plausible and is rather technical. We refer for its proof to [9], Proposition 2.1.1.

PROPOSITION

Let $f \in \mathscr{D}_{T_0^*} \cap \mathscr{D}_{S_0}$ (with no smoothness assumptions). Then $\exists$a sequence $\{f_n\}$ with $f_n \in \mathscr{D}_{T_0^*} \cap \mathscr{D}_{S_0}$ and f_n in C^1 in a neighborhood of $\bar{\Omega}$ such that as $n \to \infty$,

$$\|f_n - f\|_2 \to 0, \qquad \|T_0^* f_n - T_0^* f\|_1 \to 0, \qquad \|S_0 f_n - S_0 f\|_3 \to 0.$$

Since (8) holds when f is smooth, the proposition gives that (8) holds for all $f \in \mathscr{D}_{T_0^*} \cap \mathscr{D}_{S_0}$.

Theorem 16.2 now applies to T_0 and S_0 with $c = 1$. It follows from (4) and (5) that if $g = \Sigma_{j=1}^n g_j \, d\bar{z}_j \in H_2$, and if $S_0 g = 0$, then $\exists u$ in H_1 with $T_0 u = g$ and $\|u\|_1 \leq \|g\|_2$. Thus

$$\int_\Omega |u|^2 e^{-\phi} \, dV \leq \int_\Omega |g|^2 e^{-\phi} \, dV.$$

Now if $\Omega \subset \{z \in \mathbf{C}^n \mid |z| \leq R\}$, then

$$\begin{aligned}\int_\Omega |u|^2 \, dV &= \int_\Omega |u|^2 e^{-\phi} \cdot e^{\phi} \, dV \\ &\leq \int_\Omega |u|^2 e^{-\phi} \cdot e^{R^2} \, dV \leq e^{R^2} \int_\Omega |g|^2 e^{-\phi} \, dV \\ &\leq e^{R^2} \int_\Omega |g|^2 \, dV,\end{aligned}$$

and so (b) holds. Thus Theorem 16.1 follows from Theorem 16.3.

From now on ρ is assumed to satisfy (2) and Ω is defined by $\rho < 0$. We also shall write T and S instead of T_0 and S_0. Let us now begin the proof of (7).

LEMMA 16.4

Let $f = \Sigma_{j=1}^n f_j \, d\bar{z}_j \in C_{0,1}^1(\bar{\Omega})$. If $f \in \mathscr{D}_{T^*}$, then

$$\sum_{j=1}^n f_j \frac{\partial \rho}{\partial z_j} = 0 \text{ on } \partial\Omega, \tag{9}$$

and

$$T^* f = -\sum_{j=1}^n e^{\phi} \frac{\partial}{\partial z_j}(f_j e^{-\phi}). \tag{10}$$

Proof. Let h be a function in C^2 in a neighborhood of $\bar{\Omega}$ and $h > 0$. Put $R = h \cdot \rho$. Fix $z \in \partial\Omega$ and choose $(\xi_1, \ldots, \xi_n)$ satisfying

$$\sum_{j=1}^n \frac{\partial R}{\partial z_j}(z) \xi_j = 0. \tag{11}$$

Then at z,

$$\frac{\partial^2 R}{\partial z_j \partial \bar{z}_k} = \frac{\partial}{\partial \bar{z}_k}\left(h\frac{\partial \rho}{\partial z_j} + \frac{\partial h}{\partial z_j}\rho\right)$$

$$= \frac{\partial h}{\partial \bar{z}_k}\frac{\partial \rho}{\partial z_j} + h\frac{\partial^2 \rho}{\partial z_j \partial \bar{z}_k} + \frac{\partial^2 h}{\partial z_j \partial \bar{z}_k}\rho + \frac{\partial h}{\partial z_j}\frac{\partial \rho}{\partial \bar{z}_k}.$$

Hence

$$\sum_{j,k}\frac{\partial^2 R}{\partial z_j \partial \bar{z}_k}\xi_j\bar{\xi}_k = \left(\sum_k \frac{\partial h}{\partial \bar{z}_k}\bar{\xi}_k\right)\left(\sum_j \frac{\partial \rho}{\partial z_j}\xi_j\right)$$

$$+ h\sum_{j,k}\frac{\partial^2 \rho}{\partial z_j \partial \bar{z}_k}\xi_j\bar{\xi}_k + \rho\sum\frac{\partial^2 h}{\partial z_j \partial \bar{z}_k}\xi_j\bar{\xi}_k$$

$$+ \left(\sum_j \frac{\partial h}{\partial z_j}\xi_j\right)\left(\sum_k \frac{\partial \rho}{\partial \bar{z}_k}\bar{\xi}_k\right).$$

Now (11) implies that $\Sigma_j(\partial\rho/\partial z_j)\xi_j = 0$ on $\partial\Omega$. Also $\overline{\partial\rho/\partial z_k} = \partial\rho/\partial\bar{z}_k$, whence $\Sigma_k(\partial\rho/\partial\bar{z}_k)\bar{\xi}_k = 0$ on $\partial\Omega$. Since $\rho = 0$ on $\partial\Omega$ and $h > 0$ there, (2) implies that

(12) $$\text{On } \partial\Omega, \sum_{j,k}\frac{\partial^2 R}{\partial z_j \partial \bar{z}_k}\xi_j\bar{\xi}_k \geq 0 \qquad \text{if} \sum_j \frac{\partial R}{\partial z_j}\xi_j = 0.$$

Now choose a function h as above with $h = 1/|\text{grad } \rho|$ in a neighborhood of $\partial\Omega$. Then $R = h \cdot \rho = \rho/|\text{grad } \rho|$ there, whence $|\text{grad } R| = 1$ on $\partial\Omega$. Also Ω is defined by $R < 0$ and (12) holds.

The upshot is that we may without loss of generality suppose that $|\text{grad } \rho| = 1$ on $\partial\Omega$. It then holds that grad ρ is the outer unit normal to $\partial\Omega$ at each point of $\partial\Omega$. The divergence theorem now gives for every smooth function v on $\bar{\Omega}$,

(13) $$\int_\Omega \frac{\partial v}{\partial x_j}\,dV = \int_{\partial\Omega} v\frac{\partial \rho}{\partial x_j}\,dS$$

for all real coordinates $x_1, \ldots, x_{2n}$ in $\mathbf{C}^n$. Hence for $1 \leq j \leq n$,

(14) $$\int_\Omega \frac{\partial v}{\partial \bar{z}_j}\,dV = \int_{\partial\Omega} v\frac{\partial \rho}{\partial \bar{z}_j}\,dS.$$

Now fix $f = \Sigma_1^n f_j\,d\bar{z}_j \in C^1_{0,1}(\bar{\Omega})$, and fix $u \in C^1(\bar{\Omega})$. Then with $(\ ,\ )_j$ denoting the inner product in H_j as defined above,

$$(Tu, f)_2 = \left(\sum_j \frac{\partial u}{\partial \bar{z}_j}d\bar{z}_j, \sum_j f_j\,d\bar{z}_j\right)_2$$

$$= \int_\Omega \left(\sum_j \frac{\partial u}{\partial \bar{z}_j}\bar{f}_j\right)e^{-\phi}\,dV.$$

Fix j. Then

$$\begin{aligned}&\int_\Omega \frac{\partial u}{\partial \bar{z}_j} \bar{f}_j e^{-\phi}\, dV \\ &\quad = \int_\Omega \frac{\partial}{\partial \bar{z}_j}(u\bar{f}_j e^{-\phi})\, dV - \int_\Omega u \frac{\partial}{\partial \bar{z}_j}(\bar{f}_j e^{-\phi})\, dV \\ &\quad = \int_{\partial\Omega} u\bar{f}_j e^{-\phi} \frac{\partial \rho}{\partial \bar{z}_j}\, dS - \int_\Omega u \frac{\partial}{\partial \bar{z}_j}(\bar{f}_j e^{-\phi})\, dV,\end{aligned}$$

where we have used (14). Hence we have

$$\begin{aligned}(Tu, f)_2 &= -\int_\Omega u\left[\sum_j \frac{\partial}{\partial \bar{z}_j}(\bar{f}_j e^{-\phi})\right] dV \\ &\quad + \int_{\partial\Omega} u\left(\sum_j \bar{f}_j \frac{\partial \rho}{\partial \bar{z}_j}\right) e^{-\phi}\, dS.\end{aligned}$$

Now if $f \in \mathscr{D}_{T^*}$, it follows that we also have

$$(Tu, f)_2 = \int_\Omega u\overline{T^* f}\, e^{-\phi}\, dV.$$

Since the last two equations hold for all u in $C^1(\bar{\Omega})$, we conclude that

$$\sum_j \bar{f}_j \frac{\partial \rho}{\partial \bar{z}_j} = 0 \text{ on } \partial\Omega, \tag{15}$$

which yields (9), and that

$$\begin{aligned}\overline{T^* f e^{-\phi}} &= -\sum_j \frac{\partial}{\partial \bar{z}_j}(\bar{f}_j e^{-\phi}) \\ &= -\overline{\sum_j \frac{\partial}{\partial z_j}(f_j e^{-\phi})}, \text{ whence (10).}\end{aligned}$$

Q.E.D.

Define an operator δ_j by

$$\delta_j w = e^{\phi} \frac{\partial}{\partial z_j}(w e^{-\phi}).$$

Fix $f \in C^1_{0,1}(\bar{\Omega}) \cap \mathscr{D}_{T^*}$. By (10), $T^* f = -\Sigma_j \delta_j f_j$, and so

$$\|T^* f\|_1^2 = \sum_{j,k} \int_\Omega \delta_j f_j \cdot \overline{\delta_k f_k} e^{-\phi}\, dV. \tag{16}$$

Now fix $A, B \in C^1(\bar{\Omega})$. Applying (14) with $v = A\bar{B}e^{-\phi}$ and $j = \nu$ gives

$$\int_\Omega \frac{\partial}{\partial \bar{z}_\nu}(A\bar{B}e^{-\phi})\, dV = \int_{\partial\Omega} A\bar{B}e^{-\phi} \frac{\partial \rho}{\partial \bar{z}_\nu}\, dS.$$

Hence

$$\int_\Omega \frac{\partial A}{\partial \bar z_\nu}\bar B e^{-\phi}\,dV = -\int_\Omega A\frac{\partial}{\partial \bar z_\nu}(\bar B e^{-\phi})\,dV + \int_{\partial\Omega} A\bar B e^{-\phi}\frac{\partial\rho}{\partial \bar z_\nu}\,dS$$

$$= -\int_\Omega A\overline{\delta_\nu B}e^{-\phi}\,dV + \int_{\partial\Omega} A\bar B\frac{\partial\rho}{\partial \bar z_\nu}e^{-\phi}\,dS.$$

Writing $\int_\Omega(\ \ )$ for $\int(\ \ )e^{-\phi}\,dV$ and similarly for $\partial\Omega$, we thus have

(17) $$\int_\Omega \frac{\partial A}{\partial \bar z_\nu}\bar B = -\int_\Omega A\overline{\delta_\nu B} + \int_{\partial\Omega} A\bar B\frac{\partial\rho}{\partial \bar z_\nu}.$$

Putting $A = \delta_k w$, $B = v$, and $\nu = j$ in (17) gives

(18) $$\int_\Omega \frac{\partial}{\partial \bar z_j}(\delta_k w)\cdot\bar v = -\int_\Omega \delta_k w\cdot\overline{\delta_j v} + \int_{\partial\Omega}\delta_k w\cdot\bar v\frac{\partial\rho}{\partial \bar z_j}.$$

Direct computation gives for all u

$$\left(\delta_k\frac{\partial}{\partial \bar z_j} - \frac{\partial}{\partial \bar z_j}\delta_k\right)(u) = \frac{\partial^2\phi}{\partial \bar z_j\,\partial z_k}\cdot u,$$

so

$$-\frac{\partial}{\partial \bar z_j}(\delta_k w) = \frac{\partial^2\phi}{\partial \bar z_j\,\partial z_k}w - \delta_k\left(\frac{\partial w}{\partial \bar z_j}\right).$$

Hence

(19) $$\int_\Omega -\frac{\partial}{\partial \bar z_j}(\delta_k w)\cdot\bar v = \int_\Omega \frac{\partial^2\phi}{\partial \bar z_j\,\partial z_k}w\bar v - \int_\Omega \delta_k\left(\frac{\partial w}{\partial \bar z_j}\right)\bar v.$$

Putting $A = v$, $B = \partial w/\partial \bar z_j$, and $\nu = k$ in (17), we get

(20) $$\int_\Omega \frac{\partial v}{\partial \bar z_k}\overline{\frac{\partial w}{\partial \bar z_j}} = -\int_\Omega v\overline{\delta_k\left(\frac{\partial w}{\partial \bar z_j}\right)} + \int_{\partial\Omega} v\overline{\frac{\partial w}{\partial \bar z_j}}\frac{\partial\rho}{\partial \bar z_k},$$

which combined with the complex conjugate of (19) gives

(21) $$\overline{\int_\Omega -\frac{\partial}{\partial \bar z_j}(\delta_k w)\bar v} = \int_\Omega \frac{\partial^2\phi}{\partial z_j\,\partial \bar z_k}\bar w v - \int_\Omega v\overline{\delta_k\left(\frac{\partial w}{\partial \bar z_j}\right)}$$

$$= \int_\Omega \frac{\partial^2\phi}{\partial z_j\,\partial \bar z_k}\bar w v - \int_{\partial\Omega} v\overline{\frac{\partial w}{\partial \bar z_j}}\frac{\partial\rho}{\partial \bar z_k} + \int_\Omega \frac{\partial v}{\partial \bar z_k}\overline{\frac{\partial w}{\partial \bar z_j}}.$$

Combining (21) with the complex conjugate of (18) gives

$$\text{(22)}\qquad \int_\Omega \delta_j v \cdot \overline{\delta_k w} = \int_\Omega \frac{\partial^2\phi}{\partial z_j\,\partial \bar z_k}\bar w v + \int_\Omega \frac{\partial v}{\partial \bar z_k}\overline{\frac{\partial w}{\partial \bar z_j}}$$

$$- \int_{\partial\Omega} v\overline{\frac{\partial w}{\partial \bar z_j}}\frac{\partial \rho}{\partial \bar z_k} + \int_{\partial\Omega} \overline{\delta_k w}\cdot v\frac{\partial \rho}{\partial z_j}.$$

By (16),

$$\|T^*f\|_1^2 = \sum_{j,k}\int_\Omega \delta_j f_j \overline{\delta_k f_k},$$

so

$$\text{(23)}\qquad \|T^*f\|_1^2 = \int_\Omega \sum_{j,k}\frac{\partial^2\phi}{\partial z_j\,\partial \bar z_k} f_j \bar f_k + \int_\Omega \sum_{j,k}\frac{\partial f_j}{\partial \bar z_k}\overline{\frac{\partial f_k}{\partial \bar z_j}}$$

$$- \int_{\partial\Omega}\sum_{j,k} f_j\overline{\frac{\partial f_k}{\partial \bar z_j}}\frac{\partial \rho}{\partial \bar z_k} + \int_{\partial\Omega}\sum_{j,k}\overline{\delta_k f_k}\cdot f_j\frac{\partial \rho}{\partial z_j}.$$

Assertion.

$$-\sum_{j,k} f_j\overline{\frac{\partial f_k}{\partial \bar z_j}}\frac{\partial \rho}{\partial \bar z_k} = \sum_{j,k} f_j \bar f_k \frac{\partial^2 \rho}{\partial z_j\,\partial \bar z_k}\ \text{on}\ \partial\Omega.$$

For, by (9),

$$\sum_k f_k\frac{\partial \rho}{\partial z_k} = 0 \text{ on } \partial\Omega.$$

Hence the gradient of the function $\Sigma_k\, f_k(\partial\rho/\partial z_k)$ is a scalar multiple of grad ρ. Hence $\exists$ function λ on $\partial\Omega$ with

$$\frac{\partial}{\partial \bar z_j}\left(\sum_k f_k\frac{\partial \rho}{\partial z_k}\right) = \lambda\frac{\partial \rho}{\partial \bar z_j}, \qquad j = 1, 2, \ldots, n,$$

or

$$\sum_k \frac{\partial f_k}{\partial \bar z_j}\frac{\partial \rho}{\partial z_k} + \sum_k f_k\frac{\partial^2 \rho}{\partial \bar z_j\,\partial z_k} = \lambda\frac{\partial \rho}{\partial \bar z_j}.$$

Multiplying by $\bar f_j$ and summing over j gives

$$\sum_{j,k}\bar f_j\frac{\partial f_k}{\partial \bar z_j}\frac{\partial \rho}{\partial z_k} + \sum_{j,k}\bar f_j f_k\frac{\partial^2 \rho}{\partial \bar z_j\,\partial z_k} = \lambda\sum_j \bar f_j\frac{\partial \rho}{\partial \bar z_j}$$

$$= \lambda\overline{\sum_j f_j\frac{\partial \rho}{\partial z_j}} = 0.$$

Complex conjugation now gives the assertion. The last term on the right in (23)

$$= \int_{\partial\Omega} \left(\sum_k \overline{\delta_k f_k}\right)\left(\sum_j f_j \frac{\partial\rho}{\partial z_j}\right) = 0, \qquad \text{by (9).}$$

Equation (23) and the assertion now yield

LEMMA 16.5

Fix $f \in \mathscr{D}_{T^*} \cap C^1_{0,1}(\bar{\Omega})$. *Then*

$$\|T^*f\|_1^2 = \int_\Omega \sum_{j,k} \frac{\partial^2\phi}{\partial z_j \partial \bar{z}_k} f_j \bar{f}_k + \int_\Omega \sum_{j,k} \frac{\partial f_j}{\partial \bar{z}_k} \overline{\frac{\partial f_k}{\partial \bar{z}_j}} + \int_{\partial\Omega} \sum_{j,k} f_j \bar{f}_k \frac{\partial^2 \rho}{\partial z_j \partial \bar{z}_k}. \tag{24}$$

LEMMA 16.6

Fix $f \in \mathscr{D}_S \cap C^1_{0,1}(\bar{\Omega})$. *Then*

$$\|Sf\|_3^2 = \int_\Omega \sum_{j,k} \left|\frac{\partial f_k}{\partial \bar{z}_j}\right|^2 - \int_\Omega \sum_{j,k} \frac{\partial f_j}{\partial \bar{z}_k} \overline{\frac{\partial f_k}{\partial \bar{z}_j}}. \tag{25}$$

Proof. Since $f \in C^1_{0,1}(\bar{\Omega})$,

$$\begin{aligned} Sf = \bar{\partial} f &= \sum_\alpha \left(\sum_\beta \frac{\partial f_\alpha}{\partial \bar{z}_\beta} d\bar{z}_\beta\right) \wedge d\bar{z}_\alpha \\ &= \sum_{\alpha<\beta} \left(\frac{\partial f_\beta}{\partial \bar{z}_\alpha} - \frac{\partial f_\alpha}{\partial \bar{z}_\beta}\right) d\bar{z}_\alpha \wedge d\bar{z}_\beta. \end{aligned}$$

Hence

$$\begin{aligned} \|Sf\|_3^2 &= \int_\Omega \sum_{\alpha<\beta} \left(\frac{\partial f_\beta}{\partial \bar{z}_\alpha} - \frac{\partial f_\alpha}{\partial \bar{z}_\beta}\right)\left(\overline{\frac{\partial f_\beta}{\partial \bar{z}_\alpha}} - \overline{\frac{\partial f_\alpha}{\partial \bar{z}_\beta}}\right) \\ &= \int_\Omega \sum_{\alpha<\beta} \left|\frac{\partial f_\beta}{\partial \bar{z}_\alpha}\right|^2 + \int_\Omega \sum_{\alpha<\beta} \left|\frac{\partial f_\alpha}{\partial \bar{z}_\beta}\right|^2 \\ &\quad - \int_\Omega \sum_{\alpha<\beta} \frac{\partial f_\beta}{\partial \bar{z}_\alpha} \overline{\frac{\partial f_\alpha}{\partial \bar{z}_\beta}} - \int_\Omega \sum_{\alpha<\beta} \frac{\partial f_\alpha}{\partial \bar{z}_\beta} \overline{\frac{\partial f_\beta}{\partial \bar{z}_\alpha}}, \end{aligned}$$

which coincides with (25). Q.E.D.

Proof of Theorem 16.3. Adding equations (24) and (25) gives (7).

Note. The proof of Theorem 16.1 is now complete.

In the rest of this section we shall establish some regularity properties of solutions of the equation $\bar{\partial} u = f$, given information on f.

LEMMA 16.7

Put $B = \{z \in \mathbf{C}^n \mid |z| < 1\}$. There exists a constant K such that for $w \in C^\infty(\mathbf{C}^n)$,

$$|w(0)| \leq K\left\{\|w\|_{L^2(B)} + \sup_B\left(\max_j\left|\frac{\partial w}{\partial \bar{z}_j}\right|\right)\right\}. \tag{26}$$

Proof. It is a fact from classical potential theory that if $f \in C_0^\infty(\mathbf{R}^N)$, then

$$f(y) = C\int_{\mathbf{R}^N} \Delta f \frac{dx}{|x - y|^{N-2}}, \tag{27}$$

where C is a constant depending on N and dx is Lebesgue measure on $\mathbf{R}^N$.

Now let $\chi \in C^\infty(\mathbf{C}^n)$, supp $\chi \subset B$, and $\chi = 1$ in $|z| < \frac{1}{2}$. Then by (27) with $y = 0$ and $f = \chi w$,

$$w(0) = (\chi w)(0) = \int_{\mathbf{C}^n} \Delta(\chi w)E(x)\,dx,$$

where we put $E(x) = C/|x|^{2n-2}$. Thus

$$w(0) = I_1 + 2I_2 + I_3,$$

where

$$I_1 = \int \Delta\chi \cdot wE\,dx,$$

$$I_3 = \int \Delta w \cdot \chi E\,dx,$$

and

$$I_2 = \int (\text{grad}\,\chi, \text{grad}\,w)E\,dx.$$

With x_j the real coordinates in $\mathbf{C}^n$, we have

$$\int \chi_{x_i} w_{x_i} E\,dx = \int w_{x_i}(\chi_{x_i}E)\,dx = -\int w(\chi_{x_i}E)_{x_i}\,dx,$$

so $I_2 = -\int w\,\Sigma_i(\chi_{x_i}E)_{x_i}\,dx$.

Since χ_{x_i} and $\Delta\chi$ vanish in a neighborhood of 0 and supp $\chi \subset B$, we have, with K a constant,

$$|I_1| \leq K\|w\|_{L^2(B)}, \qquad |I_2| \leq K\|w\|_{L^2(B)}.$$

Also,

$$I_3 = \int 4\left(\sum_j \frac{\partial^2 w}{\partial z_j \partial \bar{z}_j}\right)\chi E\,dx$$

$$= 4\sum_j \int \frac{\partial}{\partial z_j}\left(\frac{\partial w}{\partial \bar{z}_j}\right)\chi E\,dx = -4\sum_j \int \frac{\partial w}{\partial \bar{z}_j}\frac{\partial}{\partial z_j}(\chi E)\,dx.$$

Since $\partial E/\partial x_j \in L^1$ locally, we have

$$|I_3| \le K \sup_B \left(\max_j \left| \frac{\partial w}{\partial \bar{z}_j} \right| \right).$$

Equation (26) follows. Q.E.D.

Choose $\chi \in C^\infty(\mathbf{C}^n)$, $\chi \ge 0$, $\chi(y) = 0$ for $|y| > 1$, and $\int \chi(y)\, dy = 1$, where we write dy for Lebesgue measure on $\mathbf{C}^n$. Put $\chi_\varepsilon(y) = (1/\varepsilon^{2n})\chi(y/\varepsilon)$. Then for every $\varepsilon > 0$,

$$\chi_\varepsilon \in C^\infty(\mathbf{C}^n), \qquad \chi_\varepsilon(y) = 0 \text{ for } |y| > \varepsilon,$$

$\int \chi_\varepsilon(y)\, dy = 1$.

Let now $u \in L^2(\mathbf{C}^n)$ and put

$$u_\varepsilon(x) = \int u(x - y)\chi_\varepsilon(y)\, dy.$$

Note that this integral converges absolutely for all x. We assert that

(28) $$u_\varepsilon \in C^\infty(\mathbf{C}^n).$$

(29) $$u_\varepsilon \to u \text{ in } L^2(\mathbf{C}^n), \qquad \text{as } \varepsilon \to 0.$$

(30) If u is continuous in a neighborhood of a closed ball, then $u_\varepsilon \to u$ uniformly on the ball.

The proofs of (28), (29), and (30) are left to the reader.

LEMMA 16.8

Let $B = \{z \in \mathbf{C}^n \,|\, |z| < 1\}$. Let $u \in L^2(B)$. Assume that for each j, $\partial u/\partial \bar{z}_j$, defined as distribution on B, is continuous. (Recall Definition 16.1.) Then u is continuous and (26) holds with $w = u$.

Proof. Fix $x \in \mathbf{C}^n$ and $r > 0$ and put $B(x, r) = \{z \in \mathbf{C}^n \,|\, |z - x| < r\}$. A linear change of variable converts (26) into

(31) $$|w(x)| \le K\left\{ r^{-n} \|w\|_{L^2(B(x,r))} + r \sup_{B(x,r)} \left(\max_j \left| \frac{\partial w}{\partial \bar{z}_j} \right| \right) \right\}.$$

Extend u to all of $\mathbf{C}^n$ by putting $u = 0$ outside B. Then $u \in L^2(\mathbf{C}^n)$. For each $\rho > 0$, put $B_\rho = \{z \,|\, |z| < \rho\}$. Fix $R < 1$ and fix $r < 1 - R$. For each $x \in B_R$, then, $B(x,r) \subset B_{R+r} = B'$.

Fix $x \in B_R$. If $\varepsilon, \varepsilon' > 0$, $u_\varepsilon - u_{\varepsilon'} \in C^\infty(\mathbf{C}^n)$. Equation (31) together with $B(x, r) \subset B'$ gives

$$|u_\varepsilon(x) - u_{\varepsilon'}(x)| \le K\left\{ r^{-n} \|u_\varepsilon - u_{\varepsilon'}\|_{L^2(B')} + r \sup_{B'} \left(\max_j \left| \frac{\partial u_\varepsilon}{\partial \bar{z}_j} - \frac{\partial u_{\varepsilon'}}{\partial \bar{z}_j} \right| \right) \right\}.$$

Now, by (29), $\|u_\varepsilon - u_{\varepsilon'}\|_{L^2(B')} \to 0$ as $\varepsilon, \varepsilon' \to 0$. Also, it is easy to see that $\partial u_\varepsilon/\partial \bar{z}_j - \partial u_{\varepsilon'}/\partial \bar{z}_j \to 0$ uniformly on B' as $\varepsilon, \varepsilon' \to 0$. Hence $u_\varepsilon(x) - u_{\varepsilon'}(x) \to 0$ uniformly for $x \in B_R$. Hence $U = \lim_{\varepsilon \to 0} u_\varepsilon$ is continuous in B_R. Also, by (29), $u_\varepsilon \to u$ in $L^2(B)$.

Hence $U = u$ and so u is continuous in B_R. It follows that u is continuous in B, as claimed.

Fix $\varepsilon > 0$ and $\rho < 1$. Then, by (31),

$$|u_\varepsilon(0)| \leq K\left\{\rho^{-n}\|u_\varepsilon\|_{L^2(B_\rho)} + \rho \sup_{B_\rho}\left(\max_j\left|\frac{\partial u_\varepsilon}{\partial \bar{z}_j}\right|\right)\right\}.$$

As $\varepsilon \to 0$, $u_\varepsilon(0) \to u(0), \|u_\varepsilon\|_{L^2(B_\rho)} \to \|u\|_{L^2(B_\rho)}$, and $\partial u_\varepsilon/\partial \bar{z}_j \to \partial u/\partial \bar{z}_j$ uniformly on B_ρ for each j. Hence

$$|u(0)| \leq K\left\{\rho^{-n}\|u\|_{L^2(B_\rho)} + \rho \sup_{B_\rho}\left(\max_j\left|\frac{\partial u}{\partial \bar{z}_j}\right|\right)\right\}.$$

Letting $\rho \to 1$, we get that (26) holds with $w = u$. Q.E.D.

LEMMA 16.9

Let Ω be a bounded domain in $\mathbf{C}^n$ and $u \in L^2(\Omega)$. Assume that for all j,

$$\frac{\partial u}{\partial \bar{z}_j} = 0 \text{ as a distribution on } \Omega. \tag{32}$$

Then $u \in H(\Omega)$.

Proof. Define $u = 0$ outside Ω. Then $u \in L^2(\mathbf{C}^n)$. By a change of variable, we get

$$u_\varepsilon(z) = \int u(\zeta)\chi_\varepsilon(z - \zeta)\, d\zeta.$$

Fix j. Note that $(\partial\{\chi_\varepsilon(z - \zeta)\}/\partial \bar{z}_j) = -(\partial\{\chi_\varepsilon(z - \zeta)\}/\partial \bar{\zeta}_j)$. Hence

$$\frac{\partial u_\varepsilon}{\partial \bar{z}_j}(z) = \int u(\zeta)\frac{\partial}{\partial \bar{z}_j}(\chi_\varepsilon(z - \zeta))\, d\zeta = -\int u(\zeta)\frac{\partial}{\partial \bar{\zeta}_j}(\chi_\varepsilon(z - \zeta))\, d\zeta.$$

Fix $z \in \Omega$ and choose $\varepsilon < \text{dist}(z, \partial\Omega)$. Put $g(\zeta) = \chi_\varepsilon(z - \zeta)$. Then supp g is a compact subset of Ω. By (32),

$$\int u(\zeta)\frac{\partial g}{\partial \bar{\zeta}_j}(\zeta)\, d\zeta = 0.$$

Thus $\partial u_\varepsilon(z)/\partial \bar{z}_j = 0$. Hence $u_\varepsilon \in H(\Omega)$.

Fix a closed ball $B' \subset \Omega$. By (32), $\partial u/\partial \bar{z}_j$ is continuous in a neighborhood of B' and so, by (30), $u_\varepsilon \to u$ uniformly in B' as $\varepsilon \to 0$. Hence $u \in H(\mathring{B}')$. So $u \in H(\Omega)$. Q.E.D.

NOTES

The fundamental result of this section, Theorem 16.1, is due to L. Hörmander. It is proved in considerably greater generality in Hörmander's paper, L^2 estimates and existence theorems for the $\bar{\partial}$-operator. We have followed the proof in that paper, restricting ourselves to (0, 1)-forms. The method of proving existence theorems for the $\bar{\partial}$-operator by means of L^2 estimates was developed by C. B. Morrey, The analytic embedding of abstract real analytic manifolds, *Ann. Math.* (2), **68** (1958), and J. J. Kohn, Harmonic integrals on strongly pseudo-convex manifolds, I and II, *Ann. Math.* (2), **78** (1963) and *Ann. Math.* (2), **79** (1964). These methods have proved to be powerful tools in many questions concerning analytic functions of several complex variables. For such applications the reader may consult, e.g., Hörmander's book *An Introduction to Complex Analysis in Several Variables* [8, Chaps. IV and V].

In Section 17 we shall apply Theorem 16.1 to a certain approximation problem.

17

MANIFOLDS WITHOUT COMPLEX TANGENTS

Let X be a compact set in $\mathbf{C}^n$ which lies on a smooth k-dimensional (real) submanifold Σ of $\mathbf{C}^n$. Assume that X is polynomially convex. Under what conditions on Σ can we conclude that $P(X) = C(X)$?

If Σ is a complex-analytic submanifold of $\mathbf{C}^n$, it does not have this property. On the other hand, the real subspace Σ_R of $\mathbf{C}^n$ does have this property. What feature of the geometry of Σ is involved?

Now fix a k-dimensional smooth submanifold Σ of an open set in $\mathbf{C}^n$, and consider a point $x \in \Sigma$. Denote by T_x the tangent space to Σ at x, viewed as a real-linear subspace of $\mathbf{C}^n$.

Definition 17.1. A *complex tangent* to Σ at x is a complex line, i.e., a complex-linear subspace of $\mathbf{C}^n$ of complex dimension 1, contained in T_x.

Note that if Σ is complex-analytic, then it has one or more complex tangents at every point, whereas Σ_R has no complex tangent whatever.

Definition 17.2. Let Ω be an open set in $\mathbf{C}^n$ and let Σ be a closed subset of Ω. Σ is called a *k-dimensional submanifold of* Ω *of class e* if for each x_0 in Σ we can find a neighborhood U of x_0 in $\mathbf{C}^n$ with the following property: There exist real-valued functions $\rho_1, \rho_2, \ldots, \rho_{2n-k}$ in $C^e(U)$ such that

$$\Sigma \cap U = \{x \in U | \rho_j(x) = 0, j = 1, 2, \ldots, 2n - k\},$$

and such that the matrix $(\partial \rho_j / \partial x_\nu)$, where $x_1, x_2, \ldots, x_{2n}$ are the real coordinates in $\mathbf{C}^n$, has rank $2n - k$.

Exercise 17.1. Let Σ, $\rho_1, \ldots, \rho_{2n-k}$, be as above and fix $x^0 \in \Sigma$. If there exists a tangent vector ξ to Σ at x^0 of the form

$$\xi = \sum_{j=1}^{n} c_j \frac{\partial}{\partial \bar{z}_j}$$

such that $\xi(\rho_\nu) = 0$, all ν, then Σ has a complex tangent at x^0.

THEOREM 17.1

Let Σ be a k-dimensional sufficiently smooth submanifold of an open set in $\mathbf{C}^n$. Assume that Σ has no complex tangents.

Let X be a compact polynomially convex subset of Σ. Then $P(X) = C(X)$.

Note 1. "Sufficiently smooth" will mean that Σ is of class e with $e > (k/2) + 1$. It is possible that class 1 would be enough to give the conclusion.

Note 2. After proving Theorem 17.1, we shall use it in Theorem 17.5 to solve a certain perturbation problem.

Sketch of Proof. To show that $P(X) = C(X)$ we need only show that $P(X)$ contains the restriction to X of every $u \in C^\infty(\mathbf{C}^n)$, since such functions are dense in $C(X)$.

Fix $u \in C^\infty(\mathbf{C}^n)$. By the Oka–Weil theorem it suffices to approximate u uniformly on X by functions defined and holomorphic in some neighborhood of X in $\mathbf{C}^n$. To this end, we shall do the following:

Step 1. Construct for each $\varepsilon > 0$ a certain neighborhood ω_ε of X in $\mathbf{C}^n$ to which Theorem 16.1 is applicable.

Step 2. Find an extension U_ε of $u|_X$ to ω_ε such that $\bar\partial U_\varepsilon$ is "small" in ω_ε.

Step 3. Using the results of Section 16, find a function V_ε in ω_ε such that $\bar\partial V_\varepsilon = \bar\partial U_\varepsilon$ in ω_ε and $\sup_X|V_\varepsilon| \to 0$ as $\varepsilon \to 0$.

Once step 3 is done, we write

$$U_\varepsilon = (U_\varepsilon - V_\varepsilon) + V_\varepsilon \text{ in } \omega_\varepsilon.$$

Then $U_\varepsilon - V_\varepsilon$ is holomorphic in ω_ε, since $\bar\partial(U_\varepsilon - V_\varepsilon) = 0$ by step 3. Since $\sup_X|V_\varepsilon| \to 0$, this holomorphic function approximates $u = U_\varepsilon$ as closely as we please on X.

Definition 17.3. Let Ω be an open set in $\mathbf{C}^n$ and fix $F \in C^2(\Omega)$. F is *plurisubharmonic* (p.s.) in Ω if

$$(1) \qquad \sum_{j,k=1}^{n} \frac{\partial^2 F}{\partial z_j \partial \bar z_k}(z)\xi_j\bar\xi_k \geq 0$$

if $z \in \Omega$ and $(\xi_1, \ldots, \xi_n) \in \mathbf{C}^n$.

F is *strongly* p.s. in Ω if the inequality in (1) is strict, except when $(\xi_1, \ldots, \xi_n) = 0$.

LEMMA 17.2

Let Σ be a submanifold of an open set in $\mathbf{C}^n$ of class 2 such that Σ has no complex tangents. Let d be the distance function to Σ; i.e., if $x \in \mathbf{C}^n$, $d(x)$ is the distance from x to Σ. Then $\exists$ a neighborhood ω of Σ such that $d^2 \in C^2(\omega)$ and d^2 is strongly p.s. in ω.

Exercise 17.2. Prove the smoothness assertion; i.e., show that d^2 is in C^2 in some neighborhood of Σ.

Proof of Lemma 17.2. Let U be a neighborhood of Σ such that $d^2 \in C^2(U)$.

Fix $z_0 \in \Sigma$. We assert that

$$(2) \qquad \sum_{j,k=1}^{n} \frac{\partial^2(d^2)}{\partial z_j \partial \bar z_k}(z_0)\xi_j\bar\xi_k > 0$$

for all $\xi = (\xi_1, \ldots, \xi_n)$ with $\xi \neq 0$.

Without loss of generality $z_0 = 0$. Let T be the tangent space to Σ at 0 and put $d(z, T) =$ distance from z to T.

***Exercise 17.3**

$$d^2(z) = d^2(z, T) + o(|z|^2). \tag{3}$$

Also

$$d^2(z, T) = H(z) + \operatorname{Re} A(z), \tag{4}$$

where $H(z) = \Sigma_{j,k=1}^n h_{jk} z_j \bar{z}_k$ is hermitean-symmetric and A is a homogeneous quadratic polynomial in z.

Equations (3) and (4) imply that

$$\sum_{j,k=1}^{n} \frac{\partial^2(d^2)}{\partial z_j \partial \bar{z}_k}(0) z_j \bar{z}_k = H(z). \tag{5}$$

Now

$$d^2(z, T) + d^2(iz, T) = 2H(z).$$

If $z \neq 0$, either z or $iz \notin T$, since by hypothesis T contains no complex line. Hence $H(z) > 0$. Because of (5), this shows that (2) holds.

It follows by continuity from (2) that

$$\sum_{j,k=1}^{n} \frac{\partial^2(d^2)}{\partial z_j \partial \bar{z}_k}(z) \xi_j \bar{\xi}_k > 0$$

for all z in some neighborhood of Σ and $\xi \neq 0$. Q.E.D.

From now on until the end of the proof of Theorem 17.1 let Σ and X be as in that theorem and let d be as in Lemma 17.2.

LEMMA 17.3

There exists an open set ω_ε in $\mathbf{C}^n$ containing X such that ω_ε is bounded and

(6) *If $z \in \omega_\varepsilon$, then $d(z) < \varepsilon$.*

(7) *If $z_0 \in X$ and $|z - z_0| < \varepsilon/2$, then $z \in \omega_\varepsilon$.*

(8) *$\exists$ a function u_ε in C^∞ in some neighborhood of $\bar{\omega}_\varepsilon$ such that ω_ε is defined by*

$$u_\varepsilon(z) < 0.$$

(9) *$u_\varepsilon = 0$ on $\partial\omega_\varepsilon$ and grad $u_\varepsilon \neq 0$ on $\partial\omega_\varepsilon$.*

(10) *u_ε is p.s. in a neighborhood of $\bar{\omega}_\varepsilon$.*

Proof. Choose ω by Lemma 17.2 so that d^2 is strongly p.s. in ω. Next choose $\beta \in C_0^\infty(\omega)$ with $\beta = 1$ in a neighborhood of X and $0 \leq \beta \leq 1$. Let Ω be an open set with compact closure such that

$$\operatorname{supp} \beta \subset \Omega \subset \bar{\Omega} \subset \omega.$$

Since d^2 is strongly p.s. in ω, we can choose $\varepsilon > 0$ such that

$$\phi = d^2 - \varepsilon^2 \beta$$

is p.s. in Ω. Further, choose ε so small that $\beta(z) = 1$ for each z whose distance from $X < \varepsilon$. Next, choose an open set Ω_1 with

$$\operatorname{supp} \beta \subset \Omega_1 \subset \bar{\Omega}_1 \subset \Omega.$$

Assertion. $\exists u \in C^\infty(\mathbf{C}^n)$ such that u is p.s. in Ω_1 and

$$|u - \phi| < \frac{\varepsilon^2}{4} \text{ on } \Omega_1. \tag{11}$$

We proceed as in the last part of Section 16. Choose $\chi \in C^\infty(\mathbf{C}^n)$, $\chi \geq 0$, $\chi(y) = 0$ for $|y| > 1$ and $\int \chi(y)\, dy = 1$. Put $\chi_\delta(y) = (1/\delta^{2n})\chi(y/\delta)$ and put

$$\phi_\delta(x) = \int \phi(x - y)\chi_\delta(y)\, dy,$$

where we have defined $\phi = 0$ outside Ω.

Then, as in Section 16, if δ is small,

$$\phi_\delta \in C^\infty(\mathbf{C}^n). \tag{12}$$

$$\phi_\delta \to \phi \text{ uniformly on } \Omega_1 \text{ as } \delta \to 0. \tag{13}$$

Also for each $(\xi_1, \ldots, \xi_n) \in \mathbf{C}^n$, $z \in \Omega_1$:

$$\sum_{j,k} \frac{\partial^2 \phi_\delta}{\partial z_j\, \partial \bar{z}_k}(z)\xi_j \bar{\xi}_k = \int \left\{ \sum_{j,k} \frac{\partial^2 \phi}{\partial z_j\, \partial \bar{z}_k}(z - y)\xi_j \bar{\xi}_k \right\} \chi_\delta(y)\, dy \geq 0,$$

since ϕ is p.s. in Ω. Hence

$$\phi_\delta \text{ is p.s. in } \Omega_1. \tag{14}$$

Choose δ such that $|\phi - \phi_\delta| < \varepsilon^2/4$ on Ω_1 and put $u = \phi_\delta$. Thus the assertion holds.

Since $u \in C^\infty(\mathbf{C}^n)$, a well-known theorem yields that the image under u of the set grad $u = 0$ has measure 0 on $\mathbf{R}$. Hence every interval on $\mathbf{R}$ contains a point t such that the level set $u = t$ fails to meet the set grad $u = 0$. Choose such a t with

$$-\tfrac{1}{2}\varepsilon^2 < t < -\tfrac{1}{4}\varepsilon^2.$$

Define

$$\omega_\varepsilon = \{x \in \Omega_1 | u(x) < t\}.$$

We claim that ω_ε has the required properties. Put

$$u_\varepsilon = u - t.$$

Then $\omega_\varepsilon = \{x \in \Omega_1 | u_\varepsilon < 0\}$. It is easily verified that $\omega_\varepsilon \subset \operatorname{supp} \beta$. It follows that $u_\varepsilon = 0$ on $\partial \omega_\varepsilon$.

Since $u = t$ on ω_ε, it follows by choice of t that grad u, and hence grad u_ε, $\neq 0$ on $\partial\omega_\varepsilon$. Thus (8) and (9) hold and (10) holds since u is p.s. in Ω_1.

Equations (6) and (7) are verified directly, using (11) and the fact that $-\varepsilon^2/2 < t < -\varepsilon^2/4$.

Thus the lemma is established. This completes step 1.

LEMMA 17.4

Fix a compact set K on Σ. Let u be a function of class C^e defined on Σ. Then $\exists$ a function U of class C^1 in $\mathbf{C}^n$ with

(a) *$U \equiv u$ on K.*

(b) *$\exists$ constant C with*

$$\left|\frac{\partial U}{\partial \bar{z}_j}(z)\right| \leq C \cdot d(z)^{e-1}, \qquad \text{all } z, j = 1, \ldots, n.$$

Proof. We first perform the extension locally.

Fix $x_0 \in \Sigma$. Choose an open set Ω in $\mathbf{C}^n$ such that $x_0 \in \Omega$, and choose real functions ρ_j such that

$$\Sigma \cap \Omega = \{x \in \Omega | \rho_1(x) = \cdots = \rho_m(x) = 0\},$$

where each ρ_j is of class C^e in Ω and such that u has an extension to $C^e(\Omega)$, again denoted u.

We assert that $\exists$ a neighborhood ω_0 of x_0 and $\exists$ integers $v_1, v_2, \ldots, v_n$ such that the vectors

$$\left(\frac{\partial \rho_{v_j}}{\partial \bar{z}_1}, \ldots, \frac{\partial \rho_{v_j}}{\partial \bar{z}_n}\right)_x, \qquad j = 1, \ldots, n$$

form a basis for $\mathbf{C}^n$ for each $x \in \omega_0$.

Put

$$\xi_v = \left(\frac{\partial \rho_v}{\partial \bar{z}_1}, \ldots, \frac{\partial \rho_v}{\partial \bar{z}_n}\right)_{x_0}, \qquad v = 1, \ldots, m.$$

Suppose that $\xi_1, \ldots, \xi_m$ fail to span $\mathbf{C}^n$. Then $\exists c = (c_1, \ldots, c_n) \neq 0$ with $\sum_{j=1}^n c_j(\partial\rho_v/\partial\bar{z}_j) = 0$, $v = 1, \ldots, m$. In other words, the tangent vector to $\mathbf{C}^n$ at x_0,

$$\sum_{j=1}^n c_j \frac{\partial}{\partial \bar{z}_j},$$

annihilates $\rho_1, \ldots, \rho_m$, and hence by Exercise 17.1 Σ has a complex tangent at x_0, which is contrary to assumption.

Hence $\xi_1, \ldots, \xi_m$ span $\mathbf{C}^n$, and so we can find $v_1, \ldots, v_n$ with $\xi_{v_1}, \ldots, \xi_{v_n}$ linearly independent. By continuity, then, the vectors

$$\left(\frac{\partial \rho_{v_j}}{\partial \bar{z}_1}, \ldots, \frac{\partial \rho_{v_j}}{\partial \bar{z}_n}\right)_x, \qquad j = 1, \ldots, n$$

are linearly independent, and so form a basis for $\mathbf{C}^n$, for all x in some neighborhood of x_0. This was the assertion.

Relabel $\rho_{v_1}, \dots, \rho_{v_n}$ to read $\rho_1, \dots, \rho_n$. Define functions $h_1, \dots, h_n$ in ω_0 by

$$\left(\frac{\partial u}{\partial \bar{z}_1}, \dots, \frac{\partial u}{\partial \bar{z}_n}\right)(x) = \sum_{i=1}^{n} h_i(x) \left(\frac{\partial \rho_i}{\partial \bar{z}_1}, \dots, \frac{\partial \rho_i}{\partial \bar{z}_n}\right)_x, \qquad x \in \omega_0.$$

Solve for $h_i(x)$. All the coefficients in this $n \times n$ system of equations are of class $e - 1$, so $h_i \in C^{e-1}(\omega_0)$. We have

$$\bar{\partial} u = \sum_{i=1}^{n} h_i \, \bar{\partial} \rho_i \text{ in } \omega_0.$$

Put $u_1 = u - \Sigma_{i=1}^{n} h_i \rho_i$. So $u_1 = u$ on Σ, and

$$\bar{\partial} u_1 = \bar{\partial} u - \sum_{i=1}^{n} h_i \, \bar{\partial} \rho_i - \sum_{i=1}^{n} \bar{\partial} h_i \cdot \rho_i = - \sum_{i=1}^{n} \bar{\partial} h_i \cdot \rho_i.$$

In the same way in which we got the h_i, we can find functions h_{ij} in $C^{e-2}(\omega_0)$ with

$$\bar{\partial} h_i = \sum_{j=1}^{n} h_{ij} \, \bar{\partial} \rho_j, \qquad i = 1, \dots, n.$$

Since $\bar{\partial} \rho_1, \dots, \bar{\partial} \rho_n$ are linearly independent at each point of ω_0, the same is true of the (0, 2)-forms $\bar{\partial} \rho_j \wedge \bar{\partial} \rho_i$ with $i < j$.

$$\begin{aligned} 0 = \bar{\partial}^2 u = \bar{\partial}\left(\sum_{i=1}^{n} h_i \, \bar{\partial} \rho_i\right) &= \sum_i \left(\sum_j h_{ij} \, \bar{\partial} \rho_j\right) \wedge \bar{\partial} \rho_i \\ &= \sum_{i<j} (h_{ij} - h_{ji}) \cdot \bar{\partial} \rho_j \wedge \bar{\partial} \rho_i. \end{aligned}$$

Hence $h_{ij} = h_{ji}$ for $i < j$. Put

$$u_2 = u_1 + \frac{1}{2!} \sum_{i,j} h_{ij} \rho_i \rho_j.$$

So $u_2 = u$ on Σ and

$$\bar{\partial} u_2 = - \sum_i \bar{\partial} h_i \cdot \rho_i + \frac{1}{2!} \sum_{i,j} \bar{\partial}(h_{ij}) \rho_i \rho_j + R,$$

where

$$\begin{aligned} R &= \frac{1}{2} \sum_{i,j} h_{ij} \rho_i \, \bar{\partial} \rho_j + \frac{1}{2} \sum_{i,j} h_{ij} \rho_j \, \bar{\partial} \rho_i \\ &= \frac{1}{2} \sum_i \bar{\partial} h_i \cdot \rho_i + \frac{1}{2} \sum_j \bar{\partial} h_j \rho_j, \end{aligned}$$

so

$$\bar{\partial} u_2 = \frac{1}{2!} \sum_{i,j} \bar{\partial} h_{ij} \cdot \rho_i \rho_j.$$

We define inductively functions h_I on ω_0, I a multiindex, by

$$\bar{\partial} h_I = \sum_{j=1}^{n} h_{Ij} \bar{\partial} \rho_j,$$

and we define functions u_N, $N = 1, 2, \ldots, e - 1$, by

$$u_N = u_{N-1} + \frac{(-1)^N}{N!} \sum_{|I|=N} h_I \rho_I,$$

where $I = (\beta_1, \ldots, \beta_n)$, $|I| = \Sigma \beta_i$, $\rho_I = \rho_1^{\beta_1} \cdots \rho_n^{\beta_n}$. Then $h_I \in C^{e-N}(\omega_0)$ if $|I| = N$, and $u_N \in C^{e-N}(\omega_0)$.

We verify

$$\bar{\partial} u_N = \frac{(-1)^N}{N!} \sum_{|I|=N} \bar{\partial} h_I \cdot \rho_I, \qquad \text{for each } N.$$

By slightly shrinking ω_0 we get a constant C such that $|\rho_I(z)| \leq C d(z)^N$ in ω_0 if $|I| = N$, and hence there is a constant C_1 with

$$\left| \frac{\partial u_N}{\partial \bar{z}_j}(z) \right| \leq C_1 \, d(z)^N, \qquad j = 1, \ldots, n, z \in \omega_0.$$

In particular, $u_{e-1} \in C^1(\omega_0)$, $u_{e-1} = u$ on Σ, and

$$\left| \frac{\partial u_{e-1}}{\partial \bar{z}_j} \right| \leq C_1 \, d(z)^{e-1}, \qquad C_1 \text{ depending on } \omega_0.$$

Also, $u = 0$ on an open subset of ω_0 implies that $u_{e-1} = 0$ there.

For each $x_0 \in K$ we now choose a neighborhood ω_{x_0} in $\mathbf{C}^n$ of the above type. Finitely many of these neighborhoods, say, $\omega_1, \ldots, \omega_g$, cover K.

Choose $\chi_1, \ldots, \chi_g \in C^\infty(\mathbf{C}^n)$ with supp $\chi_\alpha \subset \omega_\alpha$, $0 \leq \chi_\alpha \leq 1$, and $\Sigma_{\alpha=1}^g \chi_\alpha = 1$ on K.

By the above construction, applied to $\chi_\alpha u$ in place of u, choose U_α in $C^1(\omega_\alpha)$ with $U_\alpha = \chi_\alpha u$ in $\Sigma \cap \omega_\alpha$, supp $U_\alpha \subset$ supp $\chi_\alpha u$, and

$$(*) \qquad \left| \frac{\partial U_\alpha}{\partial \bar{z}_j}(z) \right| \leq C_\alpha \cdot d(z)^{e-1}, \qquad z \in \omega_\alpha, j = 1, \ldots, n.$$

Since supp $U_\alpha \subset \omega_\alpha$, we can define $U_\alpha = 0$ outside ω_α to get a C^1-function in the whole space, and $(*)$ holds for all z in $\mathbf{C}^n$.

Put $U = \Sigma_{\alpha=1}^g U_\alpha$. Then $U \in C^1(\mathbf{C}^n)$, and for $z \in K$,

$$U(z) = \sum_{\alpha=1}^{g} U_\alpha(z) = \sum_{\alpha=1}^{g} \chi_\alpha(z) u(z) = u(z) \sum_\alpha \chi_\alpha = u(z).$$

For every z,

$$\frac{\partial U}{\partial \bar{z}_j}(z) = \sum_{\alpha=1}^{g} \frac{\partial U_\alpha}{\partial \bar{z}_j}(z),$$

so, by (*),

$$\left|\frac{\partial U}{\partial \bar{z}_j}(z)\right| \leq g \cdot C \cdot d(z)^{e-1}, \qquad \text{where } C = \max_{1 \leq \alpha \leq g} C_\alpha. \qquad \text{Q.E.D.}$$

This completes step 2.

Proof of Theorem 17.1. It remains to carry out step 3.

Without loss of generality, Σ is an open subset of some smooth k-dimensional manifold Σ_1 such that the closure of Σ is a compact subset of Σ_1. It follows that the $2n$-dimensional volume of the ε-tube around Σ, i.e., $\{x \in \mathbf{C}^n | d(x) < \varepsilon\}$, $= O(\varepsilon^{2n-k})$ as $\varepsilon \to 0$.

Fix ε and choose the set ω_ε by Lemma 17.3. By (6), $\omega_\varepsilon \subset \varepsilon$-tube around Σ, so the volume of $\omega_\varepsilon = O(\varepsilon^{2n-k})$.

By (8), (9), and (10), Theorem 16.1 may be applied to ω_ε, where we take $\rho = u_\varepsilon$.

Given that u is in $C^\infty(\mathbf{C}^n)$, by Lemma 17.4 with $K = X$ we can find U_ε in $C^1(\mathbf{C}^n)$ such that for all z and j,

$$\left|\frac{\partial U_\varepsilon}{\partial \bar{z}_j}\right| \leq C d(z)^{e-1} \qquad \text{and} \qquad U_\varepsilon = u \text{ on } X.$$

By (6) this implies

$$\left|\frac{\partial U_\varepsilon}{\partial \bar{z}_j}\right| \leq C\varepsilon^{e-1} \text{ in } \omega_\varepsilon. \tag{15}$$

Put $g = \bar{\partial} U_\varepsilon$. Then $\bar{\partial} g = 0$ in ω_ε. By Theorem 16.1, $\exists V_\varepsilon$ in $L^2(\omega_\varepsilon)$ such that, as distributions, $\bar{\partial} V_\varepsilon = g$; i.e.,

$$\frac{\partial V_\varepsilon}{\partial \bar{z}_j} = \frac{\partial U_\varepsilon}{\partial \bar{z}_j}, \qquad \text{all } j, \tag{16}$$

and

$$\int_{\omega_\varepsilon} |V_\varepsilon|^2 \, dV \leq C' \int_{\omega_\varepsilon} \left(\sum_{j=1}^{n} \left|\frac{\partial U_\varepsilon}{\partial \bar{z}_j}\right|^2 \right) dV. \tag{17}$$

Equations (15) and (17) and the volume estimate on ω_ε give

$$\int_{\omega_\varepsilon} |V_\varepsilon|^2 \, dV \leq C'' \varepsilon^{2e-2+2n-k}. \tag{18}$$

By (16) and Lemma 16.8, V_ε is continuous in ω_ε. Further, fix $x \in X$ and put $B_x =$ ball of center x, radius $\varepsilon/2$. Lemma 16.8 implies that

$$|V_\varepsilon(x)| \leq K\left\{\varepsilon^{-n} \|V_\varepsilon\|_{L^2(B_x)} + \varepsilon \sup_{B_x}\left(\max_j \left|\frac{\partial V_\varepsilon}{\partial \bar{z}_j}\right|\right)\right\}. \tag{19}$$

But $B_x \subset \omega_\varepsilon$ by (7), so (18), (15), and (16) give

$$|V_\varepsilon(x)| \leq K\{\varepsilon^{e-1-(k/2)} + \varepsilon^e\}, \tag{20}$$

where K is independent of x. Thus if $e > k/2 + 1$, $\sup_X |V_\varepsilon| \to 0$ as $\varepsilon \to 0$.

Step 3 is now complete. Theorem 17.1 is thus proved.

As an application of Theorem 17.1, we consider the following problem: Let X be a compact subset of $\mathbf{C}^n$ and $f_1, \ldots, f_k$ elements of $C(X)$. Let

$$[f_1, \ldots, f_k | X]$$

denote the class of functions on X that are uniform limits on X of polynomials in $f_1, \ldots, f_k$. The Stone–Weierstrass theorem gives

$$[z_1, \ldots, z_n, \bar{z}_1, \ldots, \bar{z}_n | X] = C(X).$$

We shall prove a perturbation of this fact. Let Ω be a neighborhood of X and let $R_1, \ldots, R_n$ be complex-valued functions defined in Ω. Denote by R the vector-valued function $R = (R_1, \ldots, R_n)$.

THEOREM 17.5

Assume that $\exists k < 1$ *such that*

$$|R(z_1) - R(z_2)| \leq k|z_1 - z_2| \qquad \text{if } z_1, z_2 \in \Omega. \tag{21}$$

Assume also that each $R_j \in C^{n+2}(\Omega)$. *Then*

$$[z_1, \ldots, z_n, \bar{z}_1 + R_1, \ldots, \bar{z}_n + R_n | X] = C(X).$$

Note. Equation (21) is a condition on the Lipschitz norm of R. No such condition on the sup norm of R would suffice.

Exercise 17.4. Put X = closed unit disk in the z-plane and fix $\varepsilon > 0$. Show that $\exists$ a function Q, smooth in a neighborhood of X, with $|Q| \leq \varepsilon$ everywhere and $[z, \bar{z} + Q | X] \neq C(X)$.

Let ϕ denote the map of Ω into $\mathbf{C}^{2n}$ defined by

$$\Phi(z) = (z, \bar{z} + R(z))$$

and let Σ be the image of Ω under Φ. Evidently Σ is a submanifold of an open set in $\mathbf{C}^{2n}$ of dimension $2n$ and class $n + 2$. Since $n + 2 > (2n/2) + 1$, the condition of "sufficient smoothness" holds for Σ.

LEMMA 17.6

Σ has no complex tangents.

Proof. If Σ has a complex tangent, then $\exists$ two tangent vectors to Σ differing only by the factor i.

With $d\Phi$ denoting the differential of the map Φ, we can hence find $\xi, \eta \in \mathbf{C}^n$ different from 0 so that at some point of Ω,

$$d\Phi(\eta) = i\, d\Phi(\xi). \tag{22}$$

Let R_z denote the $n \times n$ matrix whose (j, k)th entry is $\partial R_j / \partial z_k$ and define $R_{\bar{z}}$ similarly. For any vector α in $\mathbf{C}^n$,

$$d\Phi(\alpha) = (\alpha, \bar{\alpha} + R_z \alpha + R_{\bar{z}} \bar{\alpha}).$$

Hence (22) gives

$$(\eta, \bar{\eta} + R_z\eta + R_{\bar{z}}\bar{\eta}) = i(\xi, \bar{\xi} + R_z\xi + R_{\bar{z}}\bar{\xi}).$$

It follows that $\eta = i\xi$ and

$$\bar{\xi} + R_{\bar{z}}\bar{\xi} = 0. \tag{23}$$

By Taylor's formula, for $z \in \Omega$, $\theta \in \mathbf{C}^n$, and ε real,

$$R(z + \varepsilon\theta) - R(z) = R_z\varepsilon\theta + R_{\bar{z}}\varepsilon\bar{\theta} + o(\varepsilon).$$

Applying (21) with $z_1 = z + \varepsilon\theta$, $z_2 = z$, and letting $\varepsilon \to 0$ then gives

$$|R_z\theta + R_{\bar{z}}\bar{\theta}| \leq k|\theta|. \tag{24}$$

Replacing θ by $i\theta$ gives

$$|R_z\theta - R_{\bar{z}}\bar{\theta}| \leq k|\theta|. \tag{24'}$$

Equations (24) and (24′) together give

$$|R_{\bar{z}}\bar{\theta}| \leq k|\theta| \qquad \text{for all } \theta \in \mathbf{C}^n, \tag{25}$$

and this contradicts (23). Thus Σ has no complex tangent. Q.E.D.

LEMMA 17.7

$\Phi(X)$ is a polynomially convex compact set in $\mathbf{C}^{2n}$.

Proof. Put $\mathfrak{A} = [z_1, \ldots, z_n, \bar{z}_1 + R_1, \ldots, \bar{z}_n + R_n | X]$,

$$\mathfrak{A}_1 = [z_1, \ldots, z_{2n} | X_1], \qquad \text{where } X_1 = \Phi(X).$$

The map Φ induces an isomorphism between $\mathfrak{A}$ and $\mathfrak{A}_1$. To show that X_1 is polynomially convex is equivalent to showing that every homomorphism of $\mathfrak{A}_1$ into $\mathbf{C}$ is evaluation at a point of X_1, and so to the corresponding statement about $\mathfrak{A}$ and X.

Let h be a homomorphism of $\mathfrak{A}$ into $\mathbf{C}$. Choose, by Exercise 1.2, a probability measure μ on X so that

$$h(f) = \int_X f \, d\mu, \qquad \text{all } f \in \mathfrak{A}.$$

Put $h(z_i) = \alpha_i$, $i = 1, \ldots, n$ and $\alpha = (\alpha_1, \ldots, \alpha_n)$. Choose an extension of R to a map of $\mathbf{C}^n$ to $\mathbf{C}^n$ such that (21) holds whenever $z_1, z_2 \in \mathbf{C}^n$. This can be done by a result of F. A. Valentine, A Lipschitz condition preserving extension of a vector function, *Am. J. Math.* **67** (1945).

Define for all $z \in X$,

$$f(z) = \sum_{i=1}^{n} (z_i - \alpha_i)((\bar{z}_i + R_i(z)) - (\bar{\alpha}_i + R_i(\alpha))).$$

Since z_i and $\bar{z}_i + R_i(z) \in \mathfrak{A}$ and α_i and $R_i(\alpha)$ are constants, $f \in \mathfrak{A}$. Evidently $h(f) = 0$. Also, for $z \in X$,

$$f(z) = \sum_{i=1}^{n} |z_i - \alpha_i|^2 + \sum_{i=1}^{n} (z_i - \alpha_i)(R_i(z) - R_i(\alpha)).$$

The modulus of the second sum is $\leq |z - \alpha|\,|R(z) - R(\alpha)| \leq k|z - \alpha|^2$, by (21). Hence $\operatorname{Re} f(z) \geq 0$ for all $z \in X$, and $\operatorname{Re} f(z) = 0$ implies that $z = \alpha$. Also,

$$0 = \operatorname{Re} h(f) = \int_X \operatorname{Re} f \, d\mu.$$

It follows that $\alpha \in X$ and that μ is concentrated at α. Hence h is evaluation at α, and we are done.

Proof of Theorem 17.5. We now know that $\Phi(X)$ is a polynomially convex compact subset of Σ and that Σ is a submanifold of $\mathbf{C}^{2n}$ without complex tangents. Theorem 17.1 now gives that $P(\Phi(X)) = C(\Phi(X))$, and this is the same as to say that

$$[z_1, \ldots, z_n, \bar{z}_1 + R_1, \ldots, \bar{z}_n + R_n | X] = C(X). \qquad \text{Q.E.D.}$$

NOTES

A result close to Theorem 17.1 was first announced by R. Nirenberg and R. O. Wells, Jr., Holomorphic approximation on real submanifolds of a complex manifold, *Bull. Am. Math. Soc.* **73** (1967), and a detailed proof was given by the same authors in Approximation theorems on differentiable submanifolds of a complex manifold, *Trans. Am. Math. Soc.* **142** (1969). They follow a method of proof suggested by Hörmander. A generalization of Theorem 17.1 to certain cases where complex tangents may exist was given by Hörmander and the present author in Uniform approximation on compact sets in $\mathbf{C}^n$, *Math. Scand.* **23** (1968). Theorem 17.5 is also proved in that paper, the case $n = 1$ of Theorem 17.5 having been proved earlier by the author in Approximation on a disk, *Math. Ann.* **155** (1964), under somewhat weaker hypotheses. Various other related problems are also discussed in the papers by Nirenberg and Wells and by Hörmander and the author. Further results in this area are due to M. Freeman. The proof of Lemma 17.4 is due to Nirenberg and Wells.

18

SOLUTIONS TO SOME EXERCISES

Solution to Exercise 3.2. Choose relatively prime polynomials P and Q with $Q \neq 0$ in Ω such that $f = P/Q$. For $t \in C$,

$$\frac{f(t) - f(x)}{t - x} = \frac{Q(x)P(t) - P(x)Q(t)}{Q(t)Q(x)(t - x)}$$

$$= \frac{F(x, t)}{Q(t)Q(x)},$$

where F is a polynomial in x and t,

$$= \frac{1}{Q(x)} \sum_{j=0}^{N} a_j(t)x^j,$$

where each a_j is holomorphic in Ω. Hence

$$\int_\gamma \frac{f(t) - f(x)}{t - x}\, dt = \frac{1}{Q(x)} \sum_{j=0}^{N} \left\{ \int_\gamma a_j(t)\, dt \right\} x^j = 0,$$

since each a_j is analytic inside γ. Also $\int_\gamma dt/t - x = 2\pi i$. (Why?) Hence the assertion.

Solution to Exercise 9.9. We must prove Theorem 9.7 and so we must show that $\check{S}(\mathscr{L}) \subset X$.

$\check{S}(\mathscr{L})$ is a closed subset of $\mathscr{M}$. Suppose $\exists x_0$ in $\check{S}(\mathscr{L}) \setminus X$. Choose an open neighborhood V of x_0 in $\mathscr{M}$ with $V \cap X = \varnothing$. We may assume that $\overline{V} \subset U_j$ for some j. Since $x_0 \in \check{S}(\mathscr{L})$, $\exists f \in \mathscr{L}$ with

$$\max_{\mathscr{M} \setminus V} |f| < \sup_V |f|,$$

and so

$$\max_{\partial V} |f| < \sup_V |f|.$$

Since $f \in \mathscr{L}$, $\exists f_n \in \mathfrak{A}$ with $f_n \to f$ uniformly on $\bar{V}$. Hence for large n,

$$\max_{\partial V}|f_n| < \sup_{V}|f_n|.$$

Since $V \subset \mathscr{M} \setminus X$ and $\check{S}(\mathfrak{A}) \subset X$, this contradicts Theorem 9.3. The assertion follows.

Solution to Exercise 12.2. We assert that if $R(x) = x^n + a_1 x^{n-1} + \cdots + a_n$ is a monic polynomial of degree n, then

$$(*) \qquad \max_{-1 \leqslant x \leqslant 1} |R(x)| \geq \frac{1}{2^n}.$$

For

$$R(\cos\theta) = \cos^n\theta + a_1\cos^{n-1}\theta + \cdots = \left(\frac{e^{i\theta} + e^{-i\theta}}{2}\right)^n + T,$$

where T is a trigonometric polynomial of degree $\leq n - 1$. It follows that

$$\frac{1}{2\pi}\int_{-\pi}^{\pi} R(\cos\theta)e^{-in\theta}\,d\theta = \frac{1}{2^n}.$$

Since

$$\left|\frac{1}{2\pi}\int_{-\pi}^{\pi} R(\cos\theta)e^{-in\theta}\,d\theta\right| \leq \max_{-1 \leqslant x \leqslant 1} |R(x)|,$$

we get (*).

Define a map $\phi : \mathsf{R} \to [0,2]$ as follows: Let μ represent linear measure and suppose, without loss of generality, that $\mu(S) > 0$. Put

$$\phi(x) = \frac{2}{\mu(S)}\mu\{y | y \in S, y \leq x\}, \qquad \text{all } x \in \mathsf{R}.$$

As is easily seen, then, ϕ is continuous and nondecreasing on R. Also, ϕ is constant on intervals complementary to S. Hence $\phi(S) = \phi(\mathsf{R}) = [0, 2]$. Also

$$(**) \qquad |\phi(x_1) - \phi(x_2)| \leq \frac{2}{\mu(S)}|x_2 - x_1|, \qquad \text{all } x_1 x_2.$$

Fix $x \in S$. Then, by (**),

$$|\phi(x) - \phi(\alpha_1)| \cdots |\phi(x) - \phi(\alpha_t)| \leq \left(\frac{2}{\mu(S)}\right)^t |P(x)|$$

$$\leq \left(\frac{2}{\mu(S)}\right)^t \cdot M.$$

Since $\phi(S) = [0, 2]$ this gives, for all $y \in [0, 2]$,

$$|y - \phi(\alpha_1)| \cdots |y - \phi(\alpha_t)| \leq \left(\frac{2}{\mu(S)}\right)^t \cdot M.$$

Note that (*) holds, by translation, when $[-1, 1]$ is replaced by $[0, 2]$. Apply this result to $R(y) = (y - \phi(\alpha_1)) \cdots (y - \phi(\alpha_t))$. It gives

$$\frac{1}{2^t} \leq \left(\frac{2}{\mu(S)}\right)^t \cdot M \qquad \text{or} \qquad \mu(S) \leq 4 \cdot M^{1/t}. \qquad \text{Q.E.D.}$$

Solution to Exercise 13.1. (Cf. [6, p. 107].) $f(D)$ is represented in $\mathbf{C}^n$ by a system of equations

(1) $$z_j = F_j(\lambda), \qquad j = 1, \ldots, n,$$

where the F_j are analytic in $|\lambda| < 1$ and the map: $\lambda \to (F_1(\lambda), \ldots, F_n(\lambda))$ is one-to-one. We may assume that x^0 corresponds to $\lambda = 0$; i.e., $x^0 = (F_1(0), F_2(0), \ldots, F_n(0))$. Without loss of generality, F_1 is not a constant. Hence for some $k \geq 1$ the derivative $F_1^{(k)}(0) \neq 0$ while $F_1^{(j)}(0) = 0$ for $j < k$.

If $k = 1$, F_1 is one-to-one in some neighborhood U of 0. Then $\exists$ holomorphic functions ϕ_j such that $F_j = \phi_j(F_1)$ in $U, j = 2, \ldots, n$. It follows that in some neighborhood of $x^0, f(D)$ is given by equations

$$z_2 - \phi_2(z_1) = 0, \ldots, z_n - \phi_n(z_1) = 0,$$

and so the assertion of the exercise holds.

If $k > 1$ we have to work a bit harder. We observe that $\exists G$ holomorphic and one-to-one in a neighborhood of 0 with $F_1 = G^k$. We may suppose that $F_1(0) = 0$. Then for small $\delta > 0 \, \exists$ neighborhood U of 0 such that for each ζ with $0 < |\zeta| < \delta \, \exists$ precisely k points λ_i in U with

$$F_1(\lambda_1) = F_1(\lambda_2) = \cdots = F_1(\lambda_k) = \zeta.$$

Consider $\Sigma = \{(F_1(\lambda), \ldots, F_n(\lambda)) | \lambda \in U\}$. Σ is a closed subset of the domain $\Omega: |z_1| < \delta$ in $\mathbf{C}^n$.

We shall show that Σ is an analytic subvariety of Ω.

Let h be a holomorphic function in Ω. For each ζ in $0 < |\zeta| < \delta \, \exists$ exactly k points $p_1(\zeta), \ldots, p_k(\zeta)$ in Σ with z_1-coordinate ζ. Put

(2) $$P_h(X, \zeta) = \prod_{i=1}^{k} (X - h(p_i(\zeta))).$$

The elementary symmetric functions of $h(p_1(\zeta)), \ldots, h(p_k(\zeta))$ are single-valued analytic functions of ζ in $0 < |\zeta| < \delta$, bounded for $|\zeta| < \delta/2$, and hence with removable singularity at $\zeta = 0$. Hence

$$P_h(X, \zeta) = X^k + A_1(\zeta)X^{k-1} + \cdots + A_k(\zeta),$$

the A_j being holomorphic functions in $|\zeta| < \delta$. As $\zeta \to 0$, $p_i(\zeta) \to x^0$ for each i. Hence, letting $\zeta \to 0$, we get

(3) $$X^k + A_1(0)X^{k-1} + \cdots + A_k(0) = (X - h(x^0))^k.$$

Now fix $z^0 = (z_1^0, \ldots, z_n^0) \in \Omega \setminus \Sigma$.

Assertion. $\exists H$ holomorphic in Ω with $H(z^0) \neq 0$ and $H = 0$ on Σ.

First assume that $z_1^0 \neq 0$. Then $p_1(z_1^0), \dots, p_k(z_1^0)$ are distinct points in Σ. We can hence find h holomorphic in Ω with $h(z^0) = 0$ and $h(p_i(z_1^0)) = i$, $1 \leq i \leq k$.

For $z = (z_1, \dots, z_n) \in \Omega$, put

$$H(z) = P_h(h(z), z_1) = h^k(z) + A_1(z_1)h^{k-1}(z) + \cdots + A_k(z_1).$$

Then H is holomorphic in Ω and, by (2), H vanishes on Σ.

The equation

$$P_h(X, z_1^0) = 0$$

has by choice of h the roots $1, 2, \dots, k$ and hence no other roots. In particular, 0 is not a root. Hence $A_k(z_1^0) \neq 0$. It follows, since $h(z^0) = 0$, that $H(z^0) \neq 0$, as desired.

Assume next that $z_1^0 = 0$. Choose h holomorphic in Ω with $h(z^0) = 0$, $h(x^0) \neq 0$. Again put $H(z) = P_h(h(z), z_1)$. As before, $H = 0$ on Σ. By (3),

$$A_k(0) = (-1)^k(h(x^0))^k \neq 0,$$

so $H(z^0) = A_k(0) \neq 0$, as desired.

Let $\mathscr{F}$ be the collection of all functions f holomorphic in Ω which vanish on Σ. Because of the assertion just proved, the common zero set of all functions in $\mathscr{F}$ is precisely Σ. By an elementary theorem about analytic varieties (Theorem 3 on p. 86 in [6]), that zero set is an analytic subvariety of Ω. Thus $\exists$ finitely many holomorphic functions in a ball B centered at x^0 whose common zero set in B is $\Sigma \cap B$ and hence is $f(D) \cap B$. We are done.

Solution to Exercise 13.4. Since $\phi \in P(J_0)$, $\phi(F, F_2, F_3)$ is the restriction to γ of an element ϕ^* of $\mathfrak{A}_\gamma$. Then $\phi(J_0) = \phi^*(\gamma)$, so we must show that $\phi^*(\gamma)$ is a Peano curve. We claim that for every $f \in \mathfrak{A}_\gamma$, $f(\gamma) = f(S^2)$. This will do it, for $\phi^*(S^2) \supset \phi^*(S^2 \setminus \gamma)$, which is an open set since ϕ^* is analytic on $S^2 \setminus \gamma$.

To prove the claim, suppose the contrary. Then $\exists a \in S^2 \setminus \gamma$ with $f(a) \notin f(\gamma)$. Let $z_1 = a, z_2, \dots, z_n$ be the finite zeros on S^2 of $f - f(a)$. Each $z_j \in S^2 \setminus \gamma$. Put

$$\psi(z) = \frac{f(z) - f(a)}{\prod_{j=1}^{n} (z - z_j)}.$$

ψ vanishes at infinity but nowhere else on S^2. Put

$$V(r) = \frac{1}{2\pi} \operatorname*{var\,arg}_{|z|=r} \psi.$$

For large r, $V(r) > 0$ since $\psi(\infty) = 0$. V is continuous for all r since ψ has no finite zeros. $V(0) = 0$ and V takes on only integer values. This is a contradiction, and we are done.

Solution to Exercise 17.3. Denote by $x_1, \dots, x_{2n}$ the real coordinates in $\mathbf{C}^n$. Since a rotation preserves everything of interest to us, we may assume that T is given by

$$x_1 = x_2 = \cdots = x_l = 0, \qquad l = 2n - k.$$

Since $d^2(x) \geq 0$ for all x and $d^2(0) = 0$, we have $\partial(d^2)/\partial x_j = 0$ at $x = 0$ for all j, and so

$$d^2(x) = Q(x) + o(|x|^2),$$

where $Q(x) = \Sigma_{i,j=1}^{2n} a_{ij}x_ix_j$, $a_{ij} \in \mathbf{R}$. Then

$$Q(x) = \sum_{i,j=1}^{l} a_{ij}x_ix_j + R(x),$$

$R(x)$ being a sum of terms $a_{ij}x_ix_j$ with i or $j > l$. Note that $a_{ij} = a_{ji}$, all i and j.

Assertion. $R = 0$.

We define a bilinear form [,] on $\mathbf{C}^n$ by

$$[x, y] = \sum_{i,j=1}^{2n} a_{ij}x_iy_j.$$

This form is positive semidefinite, since $[x, x] = Q(x) \geq 0$ because $d^2 \geq 0$. Also the form is symmetric, since $a_{ij} = a_{ji}$.

Fix $x^\alpha \in \mathbf{C}^n$ with $x^\alpha = (0, \ldots, 1, \ldots, 0)$, where the 1 is in the αth place and the other entries are 0. Then $[x^\alpha, x^\beta] = a_{\alpha\beta}$. If $\alpha > l$, then $x^\alpha \in T$.

If $x \in T$, then $d^2(x) = o(|x|^2)$, so $Q(x) = 0$. Fix $\alpha > l$. Then $[x^\alpha, x^\alpha] = 0$. It follows that $[x^\alpha, y] = 0$ for all $y \in \mathbf{C}^n$. (Why?) In particular, $a_{\alpha\beta} = [x^\alpha, x^\beta] = 0$ for all β. Hence $R = 0$, as claimed. Thus

(a) $$Q(x) = \sum_{i,j=1}^{l} a_{ij}x_ix_j.$$

If x is in the orthogonal complement of T and if $|x|$ is small, then the unique nearest point to x on Σ is 0, so $d^2(x) = |x|^2$. Thus if $x = (x_1, x_2, \ldots, x_l, 0, \ldots, 0)$, $d^2(x) = \Sigma_{i=1}^{l} x_i^2$, so

(b) $$Q(x) = \sum_{i=1}^{l} x_i^2.$$

Equations (a) and (b) yield that

$$Q(x) = \sum_{i=1}^{l} x_i^2$$

for all x. But $\Sigma_{i=1}^{l} x_i^2 = d^2(x, T)$. So

$$d^2(x) = d^2(x, T) + o(|x|^2).$$

Q.E.D.

BIBLIOGRAPHY

[1] E. Bishop, A minimal boundary for function algebras, *Pacific J. Math.* **9**, No. 3 (1959).

[2] E. Bishop, Holomorphic completions, analytic continuations and the interpolation of seminorms, *Ann. Math.* **78** (1963).

[3] L. de Branges, The Stone–Weierstrass theorem, *Proc. Am. Math. Soc.* **10** (1959).

[4] N. Dunford and J. T. Schwartz, *Linear Operators*, Part I, Wiley-Interscience, New York, 1958.

[5] I. M. Gelfand, Normierte Ringe, *Mat. Sb.* (N.S.) **9** (51) (1941).

[6] R. Gunning and H. Rossi, *Analytic Functions of Several Complex Variables*, Prentice-Hall, Englewood Cliffs, N.J., 1965.

[7] K. Hoffman, *Banach Spaces of Analytic Functions*, Prentice-Hall, Englewood Cliffs, N.J., 1962.

[8] L. Hörmander, *An Introduction to Complex Analysis in Several Variables*, Van Nostrand, Reinhold, New York, 1966.

[9] L. Hörmander, L^2 estimates and existence theorems for the $\bar{\partial}$ operator, *Acta Math.* **113** (1965).

[10] P. Lévy, Sur la convergence absolue des séries de Fourier, *Compositio Math.* **1** (1934).

[11] N. Wiener, *The Fourier integral and certain of its applications*, Dover, New York.

INDEX